A
LIMITLESS
MINDSET

How a highly effective leader thinks

TOM LAWRENCE

Orders: Please contact: https://highlyeffectiveleader.com, tom@highlyeffectiveleader.com

ISBN: 978-1-8383295-3-2

First published 2022

Copyright © 2022 Thomas Lawrence. All rights reserved.

All rights reserved. Apart from any permitted use under UK copyright law, no part of this publication may be reproduced or transmitted in any form or by any means, electronic or mechanical, including photocopying, recording, or any information, storage or retrieval system, without permission in writing from the publisher or under licence from the Copyright Licensing Agency Limited. Further details of such licenses (for reprographic reproduction) may be obtained from the Copyright Licensing Agency Ltd, Saffron House, 6-10 Kirby Street, London EC1N 8TS.

Printed in Great Britain

DEDICATION

To my Mum & Dad, my family, and my friends.

I love you all and I am so appreciative of all the support you have given me throughout my life.

CONTENTS

1	How Do You Improve Culture?	3
2	Influence NOT Dictatorship	15
3	Develop And Improve Yourself Every Day	27
4	It's All About Respect	41
5	Communication Is Not Enough	53
6	Inspiring Your Team To Buy-In	65
7	It's A Team Effort	77
8	Lead The Team Through Problem Solving	89
9	Unleash The Team With Leadership	101
10	Know Who Your Leaders Are	111

Do you want to improve processes and profit? You must first start with the people

CHAPTER 1
HOW DO YOU IMPROVE CULTURE?

Develop your people

When creating a leadership culture, positions don't count. People are people, leaders are leaders. However, highly effective leaders know how to create a leadership culture. Embrace it.

Becoming a highly effective leader is not a job position, and it is not part of a job description. Being a highly effective leader is a way of life, it is a lifelong commitment. Regardless of what level you are in your organisation, whether it is CEO, Director, Head of department, Manager, or Supervisor, all highly effective leaders have the same responsibility. That responsibility is to create the culture within their team that they lead, and to develop their team to live and breathe that culture.

Whatever level you are at right now on the leadership ladder, I want to help you to start thinking like a highly effective leader. You are reading this book because you want to change your mindset, and you want to either become a highly effective leader, or you want to become an even better highly effective leader for your team and organisation. If you are working on yourself every day, then I want to give you some further tools to help you develop further.

You will have the ability and the tools to lead the way in your organisation. You can create the right culture for your organisation, not just in your team, or your department. It is my job to help you to increase your influence with everyone in your organisation. When you can increase your influence, you will know that you are on the right path to becoming a highly effective leader. You will become known as "the man/woman of the people." First, when you are developing your team, they will spread your influence throughout the organisation for you.

Your influence throughout your organisation will be far and wide. The people of the organisation will make the decision to see you as their leader and will want to follow you. They will make this decision because of who you are and your character, not because of what you can do or the position you currently hold. That is when you know you have created the right culture…a leadership culture. When the organisation is led through a leadership culture, you will have free reign to create and develop more leaders. Eventually you will create highly effective team players and leaders. You do not have to be at the top of the organisation to have this kind of influence throughout the organisation. You just need the right mindset.

Most of the organisations I have worked with throughout my engineering career, whenever they wanted to make a big change, they would always bring in a consultant to help. I was never a person who liked consultants, especially when they were telling me how to lead my team when they had only been there a few days. I always thought that change could be grown from within, we just needed to work together to do it. Not be told how to manage or lead a team by a consultant who didn't have a lot of experience of our organisation.

So, whenever I worked with companies to help them, I didn't want to show the managers how to lead teams. My objective was to help them create a leadership culture that would benefit the whole organisation and help them create and develop highly effective leaders.

A leadership culture can change an organisation so much for the better. The people of the organisation are encouraged to stand out and be innovative. The level of moaning and groaning reduces a lot. There is no such thing as a blame culture anymore, and when things go wrong there is no finger pointing. In fact, when there are problems, the people are more than happy to work the problem and come to a solution together. There is less planning, and more action. There is more walking the walk than talking the talk. Most of all, the leaders of the organisation love and respect their people and their people love and respect them too.

With that respect from both the leaders and the people, everyone is a lot happier. So, when it comes to developing the people, they embrace personal growth and want to grow. They are listening to and following their leader because they want to. They are being innovative and trying to create positive change because they want to. They are trying to improve the organisation's policies and procedures because they want to. They want to do these things for the benefit of their teammates, colleagues, and the whole organisation. A leadership culture within an organisation creates a win/win situation for everyone.

I have led and created a leadership culture in places I have worked and in places I have helped. It is an amazing thing to start, be a part of, and when it is successful there is no better feeling in the world. I am encouraging you that no matter what position you hold within your organisation, have a highly effective leader's mindset, and accept the responsibility to create a leadership culture. If you do this, you will attract people who want to help you. Firstly, starting with your own team and then moving on to the next team.

This is by no means an easy task. This will be one of the most challenging things you will ever do in your life, believe me. You will need to inspire, motivate, lead, encourage, and engage with the people in your team and organisation.

I ask you, are you ready to take on this challenge?

I've been part of many organisations where they think that by putting up posters on the wall, or on notice boards with famous motivational quotes will do the trick and will change culture. Unfortunately, that is not how it works. It is also not going to work if you go around just talking about how you want to change culture, and what your vision is. To implement any kind of change, especially a culture change, we need to take a lot of action. The first action we must take is by developing the people, encouraging the people, and engaging the people. People, people, people is our first and most important port of call.

The final output of creating a leadership culture is better company policies, procedures, processes, more profit, and reduced cost. But we cannot achieve any of this without developing and helping our people.

To create a leadership culture anywhere in the world, then we must think, behave, act, and live like a highly effective leader. We are going to instigate this change of culture, so it is very important. A leadership culture can bring so many benefits to everyone involved. It becomes more than just a place of work; it becomes a way of life.

Leadership is influence, so to achieve this challenge; our influence must increase with everyone in the organisation. You can learn all about processes and management skills, but to increase your influence, you need the people, and you need to develop them.

I have attended many management skills courses and learned the tools about coaching and mentoring. But, to lead people, you need to do more than attend a few courses. You need to read books, just like you're doing now. You need to practice what you learn.

Most organisations when they are trying to change culture focus on the wrong things. They focus on the different tools you can use. For example, lean, kaizen, yamazumi etc. These are all well and good, but they focus on processes and tools rather than people. A leadership culture requires us to be deliberate on what we focus on…people. Our mission is to change culture, not to improve our processes and profit. They are a benefit and a byproduct of us changing the culture, not the focus.

What you are about to read in this book will not only help you to develop your people, change your culture into a leadership culture, increase your influence with your team and your organisation. It will help you to develop a highly effective leader's mindset and live like a highly effective leader. Leadership is a way of life, not a job. The rewards of creating a leadership culture are phenomenal, and I encourage you to grab this opportunity with both hands.

Creating a leadership culture with the people of your organisation is a team effort. They will have just as much input as you, and that is the most rewarding thing about it. Leading the people to become leaders is what increasing your influence is about. Leaders create leaders, who then create more leaders, who then create more leaders. Influence is like a snowball effect which grows and grows over time.

A compelling vision from a leader is great. But, if the leader does not act and implement the steps required to strive towards the vision, then it is not so great. There is no point in visualising and exciting your people if you don't act up on it. A leadership culture will enable the people to act upon your vision for you and make it their own…think about it.

Live your message

The difference between a highly effective leader and a low performing leader is one thing…Character!

What you will learn throughout this book is how to lead your team, grow your influence through your team, and through the organisation. My goal for you is to start thinking and feeling like a highly effective leader, how to create a leadership culture, and create an environment to allow you and your team to achieve excellent results. This book is not about management tools or management processes.

When it comes to management, we must not manage people. Only processes and things can be managed. People must be led by a highly effective leader. Leadership requires thoughts and feelings of self and others, as management does not. So, in each chapter throughout this book, you will only learn about leadership, and how to impact your own, and your team's thoughts and feelings positively.

Highly effective leaders work on themselves every single day as they know it is so important to develop themselves. You will be the toughest, and most important person you will ever lead. So, it is just as important for you to work on yourself every single day too. As you lead and develop yourself, you are respecting yourself. As you lead and develop your team, you are respecting your team, and you are increasing your influence with them daily.

By living your message, and leading your message through example, your influence will not only increase with your team, but throughout your organisation too. To really live your message, increase your influence, and create a leadership culture, you must have a high-level character. If there is something you don't know, then you cannot teach it. If there is something that you don't have, then you cannot give it away. Knowing this about yourself and being comfortable saying "I don't know" is how you build your character.

Highly effective leaders can increase their influence with anyone, not only their teams, or their colleagues. Highly effective leaders can lead teams in any industry, as the principles of leadership are the same everywhere. What I share in this book will enable you to become a highly effective leader and increase your influence with anyone. Whether it be in your current team, organisation, community, friends, or family.

To think, behave, act, and live like a highly effective leader, you do not need a fancy job title. You do not even need to be in a leadership position or have any real authority. A leadership culture does not require your people to have any of these things. However, there are lot of people who do think that you need to have the title, position, and authority, and that is why they struggle to create the leadership culture. They believe that you need to have the authority to be able to lead and achieve their desired results.

Highly effective leadership is all about who you are, and why you need a high-level character. What you are, or your job title is not what highly effective leadership is about. You do not need to have a formal position or authority to be a highly effective leader. You do need to behave, think, and live leadership.

It's all about *respect*. Having respect for your teammates, colleagues, and every person within your organisation is what is required to be a truly highly effective leader. Respect is one of the foundations of creating a leadership culture. So, as you respect every single person you will then influence others to have the same respect by following your example.

Accepting responsibility for the results your team achieve is a trait of a highly effective leader and should be adopted by everyone in the team. If you are part of the team, and when the results aren't as good as expected, if you start to blame others then you are on the path to becoming a low performing leader. Blame is not a trait of a highly effective leader and is not welcome inside an organisation with a leadership culture.

Being authentic and valuing people is most important if you want to lead your team, and your organisation through change. Especially if your change is a leadership culture and is also your message. When leading change, it goes beyond changing things in people's lives at work, it includes people's lives at home too. This change will also have a huge effect on your own life, so do not underestimate that. In my experience having tried to initiate changes at every organisation I have worked at since 2009, change is personal and affects everyone involved.

This is what it means to *live your message*. Most organisations throughout the world do not really understand what change means, and what it means to implement change. The best change anyone could ever want to implement should be focussed on people, and how we can help our people improve. To live this kind of change, we must first work on improving ourselves, and lead the change by example.

I began my engineering career when I was 16 years old as a mechanical apprentice in 1999, working for an automotive company in Liverpool. Following my apprenticeship, I completed my engineering degree, and it was then when I took my first leadership role in 2009 as a project manager. From 2009 onwards I "climbed the ladder" working as an engineering production manager, engineering performance manager, senior engineering manager, senior engineering consultant, and now I am an author.

In 2009, when I took up the role of project manager, I was working for Merseyside's train operator, and my first project was to increase the fleet of train's wheel life to six years. Without having any idea how to do it, I accepted the responsibility for this. A few of the guys who I worked with did not give me a good vibe about accepting this project, because so many people had tried this before and failed.

My boss gave me 3 months to work with the engineering team. He empowered me to lead the team in working together to come up with a solution to increase wheel life. Then when we had the solution, I was to lead the team in initiatives on how we were going to implement the solution. I worked with the engineers every single day for three months, and we came to a solution. This is what I call the beginning of my leadership journey.

Our solution was not a one off, it was making small changes to how we worked as an engineering team and changing how we did our maintenance. Basically, we worked together as a team every single day to continuously improve ourselves and our product (train wheels). We were striving to become better every day.

If we work on ourselves before we work on anything else every single day, then we will improve. We will improve in our professional lives and our personal lives. When doing this we are improving what we know (competencies), and we are improving who we are (character).

While we are working on ourselves, we are leading ourselves better, and that takes courage. Leadership is not about you, but it starts with you. When leading others, either in a team, or when at home, this also takes courage. Do you have the courage to lead yourself, and do you have the courage to lead others?

Highly effective leaders have the character to lead and influence others through change to improve themselves and improve the organisation. Low performing leaders lack character, and they do not lead their people through change well. As a highly effective leader, your team and your organisation need you to pull through when it comes to implementing new changes. It will be your character that will either pull you up or push you down.

How much do you want to succeed? How willing are you to live your message?

To succeed, we must work on ourselves every day and constantly learn. By reading this book every day you are on the right path to success.

Change requires a lot of hard work

To lead people effectively, it requires hard work, every single day. A lot of leaders do not realise just how much hard work it really is. By not realising this, they are not aware of the impact they have on their people, and don't understand why their people are so negative.

In 2009 when I took the responsibility for increasing the wheel life of Merseyside's trains, without realising it, I was living my message. Every day I was talking about increasing the wheel life, I was passionate about succeeding in this challenge, and my passion "rubbed off" on the team. The engineers I was working with did not report into me. However, I was influencing them through living my message. I was including everybody in the project, and how "we" would succeed, not "I". Leadership is influence, and you can influence anybody when you live your message.

When the team and I were working on the small changes to increase our fleet of train's wheel life, our engineering director Kevin was really pleased with what we were doing. He was so impressed, that he wanted us to increase our target to six years wheel life. This was the maximum a train wheelset could last. This would ultimately save our company a lot of money because we would not need to replace wheelsets as much. However, this was not the thinking of Kevin. The reason he wanted to stretch our target was not for financial gain, it was to stretch us as a team and as individuals. Kevin knew how to lead his people so that they could continuously improve.

Kevin was willing to give us the autonomy to do whatever it took to increase the wheel life to six years. I volunteered to continue to lead this project, and the team I had been working with wanted to continue too. So, we started by speaking with other train operators in the UK, and wanted to learn if there was anything, they were doing differently to us that helped them increase their train's wheel life to six years. The answer was to fit wheel lubricators to every train, in a fleet of 59.

The team and I put together the business case for this. Then I presented the business case to Kevin, as the leader of the project. Kevin liked what he saw and heard and signed the business case off. Within a year we sourced the supplier of the wheel lubricators, and then we fitted them to the trains ourselves. I then trained the rest of the engineering team on how to maintain and replenish the lubricators. After a few months of the lubricators being fitted, wheel life was increasing, and we eventually got to our first wheelset lasting six years. We had achieved our stretch target that was set by our engineering director Kevin.

In 2009 this was one of the most successful projects that our engineering team had undertook, and I was the leader. This was my first time leading and influencing a team of engineers who had more experience than I did. There were negative people, who thought that we would never achieve six years wheel life. But they were negative people who feared any new changes being implemented. We proved them wrong, and I was very proud of the team, and of myself.

At that time in 2009, the company I was working with didn't have any leadership development training within their engineering department. I wasn't the kind of person who would read personal growth or leadership books then either. However, Kevin was a highly effective leader who I followed because I wanted to.

The rest of the engineering team followed Kevin because they wanted to, not because they had to. If you read any kind of leadership book, he was the type of leader that these books would describe, and who you would want to follow. There are not enough leaders like Kevin around, especially in the engineering world.

Kevin encouraged me to start reading books on leadership and personal growth and put what I was learning into practice when leading the project team. Kevin knows that leadership started with him, but it was about us. He understood that leadership is a personal journey, and by letting me lead this project, he was letting me find my own path when starting my journey.

I started to buy more books on leadership and personal growth and was going through them very quickly. I loved learning about highly effective leaders of the past, and how they overcame their challenges and struggles. Leadership became a passion, and every day I wanted to express my passion and live leadership.

As I said earlier, there was a lot of negativity from other members of the engineering team. They saw what the team and I were working on and didn't like that we were trying to implement change. Kevin helped us with this. He cared about all his people in the engineering department. He was a mentor to people, he coached us, and he led from the front. But, most of all he gave us credit when we deserved it. It was never about him.

Kevin had a saying, "If you can't change the people, change the people." By that he meant, if you can't positively influence people to buy into you and your vision, then you will have no choice but to replace them. This, however, was the very last option. He would work his very hardest to implement the changes he wanted, and most of the time he would succeed. It was on very few occasions that people would resist no matter what, and he needed to bring in new people.

Even though the team and I were working well together on the wheel life project, I don't believe we would have made it as much of a success without Kevin's leadership. This was a huge change for the team, and a huge team requires a highly effective leader like Kevin.

In 2011, I moved to Scotland to work on their railway as an engineering production manager. With this role I had my own team who reported into me, and I had a lot more responsibility than before. I also continued with being a student of leadership. Reading books, taking courses, online courses, and I even did a yearlong leadership course at Edinburgh Napier University. When working with my team on shift, I tried my best to lead the team and help them to develop and improve.

When I began trying to help the people on the team develop, there was a lot of resistance. The reason for that is they were not interested in developing or improving. They just wanted to come to work, do their job and go home. That is fine, but if my team were not willing to improve, how would our results improve?

So, during a start of work briefing, I told the team, *"Our team's mission is to get results, improve on them, and improve ourselves at the same time."* I asked them, *"How are we going to do that?"* As soon as I asked that question, almost everybody on the team answered me. I asked them to work together during the day, and then come back to me at the end of the shift with their thoughts.

Three of them came back, and we had a plan. We called it a transformation plan. It was a transformation of results for the company, and for our own performance. This is when I learned, if you ever need a plan for anything, let the team do it. Do not do it for them.

Change is very difficult if you do not help your people to grow and develop. If you want to help your people, you must add value to them. To add value to them, you need to become more valuable. When you become more valuable, you can then help your team to become more valuable.

When adding value to your people, you are influencing them. When you are influencing your people, you are leading them. Highly effective leaders increase their influence with their people, every day.

There is not only one style of leadership. There are many styles, and they need to be practiced with people every day. Leadership is extremely hard work, and the earlier you realise this, the better it will be for you and for your people.

Influence comes first. Authority comes second, and second is nowhere

CHAPTER 2
INFLUENCE NOT DICTATORSHIP

Influence not authority

To lead without authority requires a person to have a lot of influence. When it comes to leadership, there is nothing more important than influence. Job titles and leadership positions do not matter. The more people who realise that the more people we will have with an increased influence.

A lot of people, and a lot of managers think that to be a leader, you must have authority. That couldn't be further from the truth. To be a highly effective leader, you MUST have influence over your people, whether they are your team, or you are leading them in a project. Authority is not required to be a true highly effective leader.

Have you ever heard your line manager or team leader say things like, *"You guys aren't listening to me, and you don't do what I tell you to do."* Or *"The team keeps missing targets."* Or *"They are not my responsibility; I can't tell them what to do"*?

I bet you have. You see, what your line manager or team leader is really saying to you or the team is, *"I have absolutely no influence over you and the team. How do I increase my influence with you?"*

What I teach and the message I try to put across to all the people I work with is, leadership is not a job title or a senior position within your organisation. Leadership is influence, and I help others to try and increase their influence. This book is a resource that will teach you how to increase your influence.

When I worked closely with teams, whether I was their formal leader or not,

I knew how to increase my influence with them. I tried to build a strong relationship with each individual and add value to them every day. By doing that I was helping them to develop, and I was giving up my time to do that. Some of the formal leaders who I worked with had zero influence because they were not willing to build those strong, meaningful relationships with their people. They made the excuse of not having the time. Really poor excuse in my opinion.

It is the leader's role to inspire their people and help them to make things happen. With a lot of the formal leaders, I worked with, they didn't believe that it was their role to make things happen with their people or their team. They had the attitude of putting their feet up and letting the team make things happen without any of the formal leader's input. They had no influence on their people anyway, so even if they wanted to make things happen, it would be very difficult for them.

I on the other hand knew how to make things happen with the team, and we made things happen together because we had strong relationships and we had influence with each other.

During the time I worked in Scotland as an engineering production manager, when on shift, I worked with different teams who didn't all report into me. So, for the engineering production team to work well, I needed to have a strong, authentic influence with them. This was difficult because the engineers who didn't report into me had their own line managers, and they didn't like me or the other production managers having an influence. There was a particular line manager who raised an issue with me when I tried to implement a new idea that would involve his team.

This line manager led the heavy maintenance team, so what they did was replace the heavier equipment on the trains when they needed overhauling. For example, motors and wheels. Anyway, the idea I wanted to help their team with was process and quality checking. We were getting complaints from senior leaders that the trains were not in as good a state as they had been previously. This was a big deal because we didn't want to get any customer complaints that could hurt our reputation.

The heavy maintenance manager did not like my idea to improve our processes and quality, in fact he wasn't bought in to any new ideas. He was a very old school kind of person.

When he raised the issue of not liking the quality and process checking idea, he complained to me that it would take too much time for his team to carry out further checks on quality after they had completed their work. Therefore, the trains would not get out on time and cause delays. He was frightened that his boss (who was also my boss) would give him a "kicking" if his team delayed the trains.

I suggested to him, *"What if we used a couple of members from another team to work alongside your team, and check the process and quality of work while they were still working?"* His reply to me was *"That wouldn't work because it would slow my team down, and because the other teams don't listen to me, it wouldn't work anyway."* This was just an excuse not to make this change, and I had heard the same excuse on a lot of other ideas I wanted to implement.

Basically, this heavy maintenance manager had no influence within his own team, or outside of his team. He was not a leader and did not know how to build relationships with anyone. All he could do was make excuses.

I asked him, *"Before we totally diminish this idea completely, why don't we give it a trial run, and let a couple of other team members work alongside your team?"* He immediately said no, but I kept on at him for about a week. I was explaining the benefits and how by doing this we could improve the quality of our trains. He gave in and agreed to do it, but he said, *"If this goes wrong, I am blaming you."* Really inspiring.

When I went to his team and discussed the idea with them, they were more than happy to trial it. I had a lot of influence with his team because I built relationships with all of them and tried to help them as best I could.

So, we did the trial for about a month and it worked well. The teams were great and worked well together. In the end, we implemented the quality and process checks permanently. The quality of the trains improved and there were hardly any delays either.

The reason the heavy maintenance manager was refusing to do this trial was because he feared change, but also, he had no influence with his team. He thought that if he went to his team and asked them to do the trial, they would refuse. Which was right, they would have. But, because I asked them, they did it. Why is that? He didn't have any influence, and I did. He didn't have a relationship with his team, and I did.

He had the authority over his team, and I had the influence. Influence wins every time.

To increase your influence with people, you must work on your character. Character is who we are, not what we do, or what we know. It is your competency that is what you know. Working on your character every day will enable you to increase your influence with everyone who encounters you.

Whether that be your team, colleagues, line managers, senior leaders, friends, or family.

When working on your own character, you must also help the team you lead work on and improve their character too. This will help you to create a leadership culture and environment within your team, and outside of your team. This happens when your team members increase their influence throughout the organisation. The leadership culture can start with you, but then organically increase through your team.

To be a highly effective leader, you must lead by example through character-based leadership. You must teach those leadership principles that you display to your team. By developing yours and your team's leadership, you are developing your character. By developing your character, you are increasing your influence. By increasing your influence, you are increasing the impact that you are making. Help your team to increase their impact, and your organisation will increase their impact too.

People say you are either a natural born leader, or you are not. WRONG! Leadership is like going the gym. To get into shape you must workout. To become a leader, you must work on your character, and add value to others…every day.

Don't plan it. Do it.

Making decisions can be difficult. It is much better to decide to do the wrong thing than decide to do nothing. However, as a highly effective leader, I know you will decide to do the right thing.

When I took my first leadership role in 2009 as a project manager, I wasn't studying leadership or reading leadership books straight away. It took me about 10 months before I took the time to start studying leadership principles. I wasn't much of a reader, and I was still doing my master's in engineering. The attitude I had then was to try and do as little work as possible in my free time, just do what I needed to get through university.

Back then, before I started studying leadership, I thought to be a leader you had to have authority and be respected by everyone. How wrong I was. It wasn't until I started reading and studying leadership that I realised that it was me "the leader" who must serve and respect the team. A lot of the leaders I worked with were good, but they were not students of leadership. People followed them, but mostly because they had to.

I've now been a student of leadership and personal growth for 11 years. I have read lots, and lots of leadership books, took online courses, took physical courses, workshops, spoke on the subject, and I have written my own E-book on Leadership and Influence, and personal growth. Over this time, I have led teams and mentored individuals, and I believe I am the right person to help you become a highly effective leader.

By writing this and my other books and telling you stories that happened while on my leadership journey (which I am still on), my motivation and purpose is to inspire you. I want to motivate you to view leadership and personal growth differently than you did before. My goal for you after you have read this book, is to take leadership and personal growth a lot more seriously, and work on these areas in your life every day.

How much do you respect your team? How much do they respect you in return? Do you believe that you should respect your team? Or do you believe that it is you who should be respected only?

Previously we discussed character, and that you must lead by example through character-based leadership. It is how much character you have that determines how much influence you have. It is how much influence you have that will determine results. Competency plays a part in results, but only very little. That is why we need to work on our character every single day, because character doubles our competency.

With character and influence comes respect. Respect for the people you lead is the beginning of how you influence the people you lead. The more respect you have for your people, the more you can influence them. However, respect starts with your character.

You learned earlier how I implemented a quality control and process check idea through building relationships with a team I was not the line manager of. The following example is a story of how I worked on my character, and by doing that how I overcame quite a big obstacle. In this story, I needed to control myself a lot more.

If you can control yourself in any situation then that is the right thing to do. If you are doing the right thing, then you are doing the best thing.

In 2019 when I was a senior engineering consultant, I took on a project to deliver an overhaul of two London Underground fleets. This would include leading a team of nine engineers, and a budget of £870k. The duration of this project would be for two years, and I would be working alongside leaders from the London Underground company.

When the project started, I met with the London Underground leaders and discussed how this project would go. There were a lot of options on the table, mainly on where we would be based, what days we would visit the engineering depots, and who would be doing what task and when. These were only options at the time, and nothing was confirmed. So, we agreed to keep these options confidential until we had a confirmation.

The week later, I met with the nine engineers I would be leading. Four of them were contractors and the other five were permanent employees and worked for the same company I did. It was good to have a mix of contractors and permanent because I wanted to the contractors to share their experiences as they had worked in a lot more places than the rest of the team (including me).

I arranged one to one's with the team first so I could start to build relationships from the very start. My first one to one was with a contractor called John.

My first question was, *"How are you feeling about working with us and London Underground on this great project?"*

His reply to me was, *"If you think I am going to be sitting behind a desk all day looking at spreadsheets, you've got another thing coming. I don't think you've ever managed a project like this before, have you?"*

I couldn't believe it. This was my first encounter with this guy. I was trying to break the ice and he was so angry right from the beginning of the conversation.

I knew I had to respond to him in the right way. I couldn't reply to him in an angry way too and ask him who he thinks he's talking to. That would have made it a lot worse. So, I just remained calm and controlled, and I asked, *"Where is all this anger coming from, and why do you think you will be sitting behind a desk all day looking at spreadsheets?"*

That initial response to him didn't make him any calmer, in fact he became a bit angrier. But I knew that if I became angry then there would be no way of building a relationship and increasing my influence with John. To be a highly effective leader, then self-control is paramount when it comes to confrontations like this. Especially if the other person cannot control themselves.

Anyway, the conversation continued, and I asked him some probing questions to try and find out why he was so angry, and why he thought he would be looking at spreadsheets all day. The conclusion was that he had spoken to one of the London Underground people, and they had heard a rumour from my initial conversation them that my team will be office based, and there wouldn't be a need for us to do any real engineering.

So, when we said to each other that we will keep our options confidential, someone (and it wasn't me) didn't keep to their word, or he didn't listen very well.

I told John that the reason we wanted him in my team is because he had the most experience of working as a contractor with London Underground, and that his engineering ability was excellent. I assured him that we needed him most. I promised him that the rumour he had heard was untrue. I said to him that if London Underground is a place for rumours then I would want to leave and work with someone else. There was no respect for the people if one of their leaders was spreading a rumour.

While carrying on this conversation, John became a lot calmer, and we began hitting it off.

The next day, John's attitude was great, and he helped me start the planning process for the project.

During my career and lifetime, I was not always the best at controlling myself in situations like this. If someone was angry towards me then my reaction would be to get angry back at them and get nowhere. However, by studying leadership and personal growth, I learned self-control through building my character every single day and having respect for everyone I encountered. It is a highly effective leader's character is how they resolve situations and overcome obstacles.

There are lots, and lots, and lots of excuses why a person cannot do what they want to do. However, to do the thing they want to do, they only need one reason.

You can influence anybody

To achieve the right results, it takes discipline, self-control, and teamwork. These are all decisions that you make. Are you willing to make them?

When you are asked to do something, whether that be at work or at home, do you do the bare minimum or go beyond what is required?

How often do you do more than is required of you? How often do you better than is required of you? How often do you do things before they are required of you?

I ask these questions because to do more, better, and before they are required of you is not easy. You need discipline, and you need self-control. If you need discipline, then you must decide to have discipline. If you need self-control, then you must decide to have self-control. These are decisions just like when you decide to get up in the morning, or what time you decide to eat dinner in the evening.

It is also a decision to embrace personal growth, and deliberately work on yourself to develop your character. Have you made this decision yet?

I talk a lot in my books and materials about building relationships with your team, and your peers. But we must also build relationships with the leaders who sit above us in the organisational chart. This can also be our parents and grandparents when we are at home.

For us to influence our leaders in a positive way then we must decide to build relationships with them. The best way to do that is to accept the responsibility for achieving results for the organisation, no matter what the circumstance or situation is. Then whatever kind of results our leaders want us to achieve, we go out and achieve those results for them and the organisation. Do not try to achieve the results that only you decide to achieve. That will decrease your influence with your leader.

The most highly effective leaders throughout the world are the ones who achieve results for everyone they are involved with. No matter what the circumstance or situation. This includes the people who tear into you because they don't want to *"look at spreadsheets all day."* Even the people who put you down or talk about you behind your back. Even the low performing leaders who don't believe it should be them working hard to achieve results. We must do our best and be our best to achieve results for everyone we are involved with…no matter what!

When I worked in London as an engineering technical manager, I was at mid-senior level. I also had the role of deputy head of engineering at the same time. During my time in this role, I was asked along with the other mid-senior engineering managers to form a team, and work on a project called "Maintenance, Planning, Resource, Model" or "MPRM" for short. It was a great opportunity for all of us to raise our profiles within the organisation. But, most importantly it was a great opportunity for us to improve our maintenance processes, improve how we planned our maintenance, and develop our people.

To begin this project, we met in a hotel to get away from the day-to-day business. The head of engineering and head of production met with us too (our bosses). They explained exactly what the purpose of this project was, and the results they wanted to see. However, they wanted the "how we do it" to be down to us. They did not tell us "what to do" or "how to do it", just what they wanted to see as a result.

This was music to all our ears because they made us feel like senior leaders. It felt like they were handing the baton over to us. We were now taking the reins and leading the engineering department. So, when they had given us the project scope and told us what results they were seeking, they left the hotel, and it was now up to us.

I was head of the engineering technical department, so anything that had to do with my department, I had responsibility for. I was happy to accept this responsibility because I knew that I could get things done and make things happen. I also knew I had the support of the MPRM team. As I say, it was an excellent opportunity for myself and the MPRM team to achieve results that were very, very important to our senior leaders. I knew I could contribute to making our maintenance and planning of maintenance more efficient. I also knew I could help develop my people in my department.

We decided that the best way to do this would be to set up an event and call it the "MPRM event" (or something like that). We would invite everyone from the different engineering departments. That would be technical, production, planning, materials, admin, basically the whole of engineering.

We set up the event in a hotel to get everybody away from the day-to-day as we had done with our project kick off meeting.

Everything we had done so far was planned and executed well. Our senior leaders were really pleased with us.

The MPRM event was a two-day event to ensure we got everyone there. Every person on the MPRM team had to give a presentation to the group. We presented on our own departments and what the benefits of MPRM would give us. I focussed on the people development part of the project because I knew that's what they wanted to know most about. The maintenance and planning side was good, but for me it wasn't the most important. I had the belief that if you help to develop your people, then everything else would take care of itself.

Everybody in the group on both days really enjoyed my presentation and were excited for MPRM had to offer us as an engineering team.

Following the presentations and discussions, we had a social evening so we could relax and talk about the project on a one-to-one basis with people. I soon realised that the head of engineering (my boss) and the head of production were not happy with my presentation. They were fuming. They took me aside and told me that I focussed on the wrong thing. I should be focussing on making our maintenance and planning more efficient, and then we would develop our people later.

I was "told" that from then on, start my side of the project focussing on maintenance, not development. So, I did as I was "told".

The MPRM team would meet on a weekly basis and we would help each other set up our tasks for the week. I am the engineering technical manager, and my first task was to work out how to make the maintenance instructions, and technical documents simpler. The phrase I used was *"write them so your mother can understand them."* This was supposed to be a joke, but everybody loved the phrase, and that I should lead this task with that approach. This I thought would undermine our engineers and would make it seem that they are not capable of understanding technical documents.

When I discussed this with the head of engineering, he agreed that I should take the *"write them so your mother can understand them"* approach. So again, that's what I did.

Looking ahead to about a month later. I had worked with my technical team every day on making the maintenance instructions a lot simpler, and we had done it. We trialed some maintenance tasks using the new, simpler instructions and the engineers preferred it. The task was completed in a quicker time than before, and the quality was a lot better. Wow, I thought. After me thinking that it would undermine the engineers, it made them better at their job. The results that head of engineering and head of production wanted; my team was achieving.

I learned very quickly then that I needed to listen to others more. Especially my senior leaders and my own team.

So, as I said earlier, if you want to have a positive influence on your senior leaders, then do everything you can to achieve the results that they want to see. No matter what the circumstance is, do your best for them, the team, and the organisation. Make it happen!

A LIMITLESS MINDSET

Don't keep telling people what you are going to do. Do it and surprise them.

You must develop yourself before you can develop others

CHAPTER 3
DEVELOP AND IMPROVE YOURSELF EVERY DAY

Leadership is about others, but it starts with you

How often to do you work on yourself to develop your leadership and personal growth? Do you do more than is expected? Do you more than you are paid for? If you do, then you are well on your way to becoming a highly effective leader.

As an engineer and then developing into an engineering leader from 1999 to 2009, most of the work I did on myself happened when I was at work. Or, in other words, most of my personal development happened when I was being paid to do it. That is the case with a lot of developing engineers/managers in my industry. It is probably the same in yours. It wasn't until 2011 when I moved to Scotland to become an engineering production manager, that I started working on myself outside of work time.

Highly effective leaders work on themselves with purpose, and they purposefully do it both inside of work and outside of work. They work on their leadership development and personal growth, and they do this every day, so it becomes a habit. I have personal experience that making it a habit to work on your leadership development and personal growth daily, can be amazing for your life and career. That is what I want to help you and many others do, so you can become highly effective leaders.

It wasn't until I started to work on myself with purpose that I discovered my passion for leadership development and personal growth. I could feel myself getting better as a leader, and I could see the teams I was leading getting better too. I was reading leadership books and sharing what I had learned. When I worked for Siemens in 2015, I put together leadership videos that I would upload onto the Siemens social network (they are still there today). When I did this, I received a lot of messages from people all over the world thanking me. I even received an award for doing it. Nobody had asked me to make the videos, I was just so passionate about the subject and I wanted to help people.

The more videos I made, the more people I helped, and the more people I helped, the more videos I made. It was helping people that was my real passion. That it when I learned: leadership starts with me, but it is not about me. It is about others and helping others.

Throughout my career, I received lots of "leadership training", but the training seemed to be more about the business we were working for, rather than the people we worked with. I was never taught that leadership starts with us, but it was about others. I was never taught that leadership was about serving others, and not the other way around. I was never taught that leadership is influence, and nothing else.

I found out that leadership is about others, about serving others, and that leadership is influence when I worked on myself and did my own leadership development.

My own discoveries about leadership and personal growth and sharing them with the teams I led in my career is the reason I have written other books on these topics. I want to share my knowledge on a broader scale and influence as many people as possible. I want to help you discover your very own leadership and personal growth path and set you on your journey.

You couldn't imagine just how much content is already available to you that will help you with your leadership development and personal growth. But, because most people only focus on their personal development during work time, they do not make the discovery of this content for themselves. The content they are mainly exposed to is what their companies supply to them. In my experience this content is not focussed on the right things. I want to help change that.

It wasn't until 2009 when I worked in Liverpool for their train operator that the engineering team received any kind of leadership training. As I have said earlier, it wasn't the best training but at least it was something. Before 2009, the leaders I reported into hadn't received any kind of training. They were just doing the best job they could to lead/manage us. Most of the time they did a great job, and they became good through learning from their own mistakes.

The training we started to receive in 2009 focussed on how to get the best results for the train operator. It was a three day away day in a hotel and I think we focussed on people for half a day. The rest of the time we focussed on trains, and how we could reduce delays in service. When really, we should have focussed on trains for half a day, and the rest of the time focussing on our people. But we didn't know what we didn't know, and I didn't know what I know now.

Organisations who provide leadership development to their employees must include personal growth and teach exactly what leadership is all about.... others.

But that doesn't mean that you just rely on your organisation's leadership development training. In fact, far from it. You must work on yourself every day with purpose. When working on yourself, focus in the areas of leadership development and personal growth. The material and content that is at your fingertips are: leadership books, leadership blogs, leadership you tube videos, and lots more. Whatever platform you prefer to consume this content is up to you. Just make yourself a promise that you will do it, as this really matters.

Whatever industry you currently work in, or want to move into in the future, the leadership principles are all the same. What you should be learning are the leadership techniques that will help you no matter what circumstance or situation you are in. A lot of people only focus on and practice principles that will only help them in one certain circumstance. We need to realise that leadership is not textbook, it is a timeless principle and there are many of them. Leadership is difficult and cannot be learned overnight. It is a lifelong journey of constant learning.

When I started to work on my leadership development and learning the different leadership principles, I thought to myself that a lot of the principles wouldn't work in my team, or my industry. The reason for that was because I had not yet been in a situation where this principle was required. So, I didn't learn about it. For example, when I was first confronted by one of my team members for the way I spoke to them in 2011. I didn't know how to handle the confrontation, which was totally justified by the way. So, I confronted him back and changed the subject to the work he had done that day and that I was disappointed in him.

This was totally the wrong thing to do on my behalf. I made one of my team feel even worse and caused even more stress for him. It was my role to help him, not hurt him. The principle I should have used was listening. I should have listened to what was on his mind, if I agreed then I should have apologised and worked out how to move forward in a win-win for both of us.

From then on, I vowed to myself that I would constantly learn about all leadership principles. This is lifelong learning and there is no way I will become an expert in all of them. But I know how to handle almost every situation a lot better than I could. My recommendation to you is to read at least thirty minutes a day (an hour really) on leadership development. Or watch leadership videos that last up to thirty minutes (an hour really). Do this every day and you will be so thankful you did it, trust me.

What you are doing is investing time every day in yourself. Investing in yourself requires discipline and sacrifice. Do you have the discipline, and what are you willing to sacrifice?

Remember, work on yourself with purpose.

Nobody in this world is a "natural born leader." I truly believe that everybody is born with the potential to become a highly effective leader. The people who become highly effective leaders, do so because they work on themselves every day, and develop the potential they were born with. The low performing leaders, or the people who do not want to become leaders do not work on themselves. So, therefore their potential is wasted. It is your decision. Do you want to become a highly effective leader, help others to develop their potential and follow you in becoming a highly effective leader, or do you want to stay where you are right now?

Highly effective leaders develop their team and others

Highly effective leaders know the leadership process. They know how to take their team through this process, and they know that the process is slow. To take each individual member of your team through this process, you must understand them and respect them.

In the company you currently for, or the business that you run, do you get buy-in from your people when you implement new change, or new ideas? If you do, then well done and keep going. If you don't, do you know why? Or are you still trying to figure out why? The answer to this is…your people do not feel that you or your leaders respect them. When I say respect, I don't mean letting them be involved in projects outside of their day-to-day operations. I don't mean giving them a twice a year objective review and then giving them a 2% pay rise. That is not respecting your people.

When leaders are giving 2% pay rises and allowing their team members to work on different projects, they are doing their best to keep their people happy. They are trying their hardest to stop their people from leaving the company. Highly effective leaders, and companies that are led by highly effective leaders do something different. They respect their people whole heartedly. How do they do this? When it comes to improving policies, procedures, and processes, they listen to their people's ideas first. They take them onboard and try to implement the best ones. But most of all, they show their people respect by deliberately developing them, and helping them in their personal growth.

So, when you go back to work and are with your team, what are you going to focus on? Are you going to focus on how to get better results? Are you going to focus on the numbers? Are you going to focus on how to make your leaders happier? Or are you going to focus on having respect for your people? The decision is yours. I hope you make the right one for you, your team, and your organisation.

When I talked about my engineering director Kevin earlier, he was a leader that respected his people. He respected me by deliberately developing me. So, from that development, and respect, how did I repay him? I worked a lot harder to get the best results on my projects. I went the extra mile by working more hours. I helped with projects outside of my area. I did what I could to help him and the team when developing others.

So, when you focus on developing your people, and showing them the respect, they deserve, your people will start to feel different. When they feel different about you and the company, they will go the extra mile. They will do their best to improve results in their area of the business. They will do their best to help others where they can. They will talk about you and the organisation they work for in a more positive way (word-of-mouth advertising). They will live and share the values that you have shared with them. In other words, you will have put them on the right path to becoming leaders.

However, even though these things sound great and these changes can and will happen, be extremely careful not to go about developing your people in the wrong way. By that I mean, make sure you focus on the person or people. Do not focus on the result of, going the extra mile, more productivity, more word-of-mouth advertising. These results are a natural reaction and by-product from the development work you do *with* your people. Like I said earlier, "*Leadership is not a race, it will not happen overnight. It is a slow process; it is a lifelong journey*". Do not carry out development *to* your people. Carry out development *with* your people. Your people will know the difference between having development *done* to them and having development *for* them. The feelings of both are much different.

When people *feel* that you are carrying out development work *to* them, the natural reaction or by-product of that is negativity, or refusal to accept what you think you're trying to do. Some people can react with hostility. For example, a colleague of mine in 2001 head-butted his line manager because he felt manipulated. He was fired on the spot as you can imagine.

Another example is when I went on the three-day leadership course in 2009, which we discussed earlier. When we were on the course, we felt motivated, everything sounded great (even though it wasn't) and after the three days, we were new leaders. But that couldn't have been further from the truth. The day we set foot back into the maintenance depot, everything went back to the way it was. Why was that? We didn't sustain the motivation, and we were not encouraged by our leaders to implement any new ideas. Even the leaders went back to the way things were. So, what a total waste of time and money.

Our leaders didn't realise that when you learn something new, and are motivated by it, you must keep working at it. They thought that now the three days are over, these new skills we have learned will naturally happen to us. No, we needed to keep working at them, and on them, every single day. We needed to work on ourselves with purpose, every single day.

Those leaders in 2009 were low performing leaders and had a lot to learn if they were to ever become highly effective leaders.

Low performing leaders have the exact same mentality as low performing team members. They will do just the bare minimum required so that they can satisfy their boss and ensure that they have a job the next day. I don't know about you, but that is very sad, and totally unfair on the team they are "leading." Their people only follow them because they must, not because they want to.

Highly effective leaders show their team respect. They develop their team correctly, and they also encourage their team to work on themselves every single day. If they send their team on a three-day course, when the team come back, the leader will sit with each individual and discuss what they learned. Then they will help everyone develop a daily action plan so that they can work on their newly found skills and keep developing them with purpose. Their team members follow them because they want to. They show their leader respect because they feel respected by their leader. The leader and the team work together to help themselves and the team succeed.

I have always done my best to help develop my teams and the people who I mentor in leadership and personal growth. I have either taught the principles myself, introduced them to books they can learn from, videos on you tube, and other materials. So that they can use these materials to work on themselves every day and develop.

For you to become a highly effective leader, you must go beyond showing your team the ropes to get the job done. You must go beyond the 2% pay rise or the offer of working on different project. You must show them the respect by helping them to develop. With your current team, or any team you lead in the future, accept the responsibility for being the person that is going to help your people grow and develop.

This might seem strange to you, but I started taking some of the teams I worked with through a leadership mastermind group. The topic to be discussed in the mastermind group was a book called *"21 irrefutable laws of leadership"* written by John Maxwell. Each law will help you in whatever situation you come by on your leadership journey. I would strongly recommend on you getting a copy of this book.

Anyway, during the mastermind group everybody on the team went and read the book, and then when we came together, we discussed it as one mastermind. Then, following our mastermind, we came out of it feeling like we were better people. We turned some of the laws into our own team values that we lived every day. So, is this something that you think you could do with your team?

Having respect for your people is taking your people through the slow process of leadership and personal growth. It is not doing something to them quickly, just so you can say that you have developed your people and you will get a pat on the back. Do not fall into that trap.

Have respect for your people, develop them, care for them, and they will do the same for you.

Leadership and personal growth do not have a stopping point. Nobody in the world is a complete Master of Leadership. It is an ever-changing process. So, what worked well for you last year in your team, may not work as well now. So, we need to be constantly working on ourselves, and developing our teams to maintain and exceed our current level.

Character over competency

When we don't develop ourselves, we are damaging ourselves. When we don't develop our teams, we are damaging our teams. Why don't we develop ourselves and others?

There are many styles of leadership, and different leadership principles that I discuss throughout my website and my other materials. One thing I have learned over the years is, there is not one size fits all. In other words, every individual and team are different. Every circumstance and situation are different. Just because one principle or style worked with one person, doesn't mean it is guaranteed to work with the next person. It takes hard work to understand our people, and how we can help them and add value to them.

I have worked with many leaders who said that they made a success of themselves in a previous leadership role. So, they came to work with us thinking they could behave, lead, and treat us in the same way, and it would be another success. How they were wrong. They didn't understand that there was not one size leadership style fits all. They thought their new leadership role with us would be a breeze, when it was much, much harder than they thought. Some of these leaders learned from their mistakes, and did their best to work hard with us, and help us. Others were too stubborn, and in the end were removed from their position because they were making things worse.

Organisations are far too focussed on results when they should be focussing on their people and their culture. When organisations bring in new leaders, their main objective and priority is to improve the results. I am not saying we shouldn't focus on results, but they should not be the priority. The people should be the priority. Protecting the culture and living the company values should be another priority of the leaders. Having respect for the people should be another priority. Without respect for the people, there is no way of protecting, or keeping the right culture and environment for the people to work and thrive.

Focussing on processes and results will damage the culture, will damage the environment, and most of all damage the relationship with the people.

The leaders who were stubborn were not willing to let go of what they had learned about leadership, unlearn it, and then relearn new leadership principles that would help them in their situation. They were failing in their leadership of us and their other teams but were not learning from their failings. To unlearn, and then relearn something new, that is the mindset of a highly effective leader. You need a strong mindset and be willing to accept and admit that you don't know it all. We need to be learning every day. That is why highly effective leaders work on themselves every day.

The mindset of a highly effective leader is to be constantly improving. Their thinking is, if I improve myself, I am making myself more valuable to my team. By being more valuable to my team, I am in a great place to help them become more valuable too.

How valuable to your team are you right now? Do you feel that you are adding value to your people? Are you willing to challenge yourself, and work on yourself every day to become more valuable?

There are two principles I would like you to focus on when working on yourself:

1. Continuous improvement
2. Respect for your people

You should focus on continuous improvement 20% of the time and focus on having respect for your people 80% of the time.

Respect for your people is the most important principle, which is why we focus on it the most. It is the foundation of leadership and will support every other leadership principle that we study, practice, and improve. Highly effective leaders think about respect for the people more often than they think about anything else. They spend 80% of their time thinking about how to respect their people.

When highly effective leaders focus on continuous improvement, they are working on themselves to become more valuable. By becoming more valuable, they will continuously improve on the value they can add to their team. Continuously improving on their value will in turn make their team more valuable. Adding value to their team and helping them become more valuable is how to respect their team. When the team feel respected, they will improve. If they feel disrespected, they will not improve. Highly effective leaders spend 20% of their time thinking about continuously improving.

When leaders want to implement new ideas, or bring in a new change, it is imperative that they work hard on influencing their people to buy-in to the new idea or change. What is the best way for them to achieve that? By respecting the people. If the leader has respect for his/her people, the people will respect them in return, and will buy-in to the new change. If the leader dis-respects his/her people, the people will dis-respect them in return, and won't buy-in to the new change.

When highly effective leaders want to increase their influence with their team to create buy-in, they must first work on themselves to develop their character, and competency. They focus on developing their character 80% of the time, and they focus on developing their competency 20% of the time. Developing character is how a highly effective leaders improves how they lead their team, and how much they respect them. Developing competency is how a highly effective leader improves how they manage the processes and policies they have accepted the responsibility for. Leadership is focussed on people. Management is focussed on processes, policies, and things.

Most organisations provide the training for the people so that they can improve and develop their competency. This is so they can carry out the day-to-day job they were employed to do. However, they don't provide the leadership development training so that they can improve and develop their character. Therefore, you have highly effective managers rather than highly effective leaders.

The leaders of these organisations focus on competency over character. When focussing on competency development more than character development, they are focussing 80 - 90% of their time on continuous improvement, and 10-20% of their time on character development. May be even less time on character development. They have it the wrong way around.

The leaders of these organisations don't even think about developing their own character, never mind the people in their charge. It's an unfortunate fact, but these organisations are prioritising and focussing on the wrong things. This needs to change.

Low performing leaders will tell their people that they are developing them, but they aren't. They manipulate their people to get what they want by telling them, *"This task is part of your development."* When, it is a task that they have done many times before or has nothing to do with what their team is about. They manipulate their team for very selfish reasons. They do not think about how their team thinks or feels.

To be a highly effective leader, you must increase your influence with your team. To increase your influence, you must develop your character, and you must also develop the character of each of your team members, so that they can increase their influence. You must also help your team to develop their competency so that the processes improve along with the people. To be a highly effective leader, you must become a highly effective leader. Focus 80% on character development, 20% on competency development.

The more character you develop, the more respect you will have for your people. The more respect you have for your people, the more character your people will have. Remember, to grow your people, you must grow yourself.

What is your business about? People, things, or results? Focus on the things, you are not developing your people, and not producing results. Focus on the results, you are not developing the people, and nothing will change. Focus on the people, and the things will become easier, and results will get better.

To lead others through change; you must first lead yourself through change

CHAPTER 4
IT'S ALL ABOUT RESPECT

Experiencing the change

Taking a team or an organisation through change is an experience. However, not many see it like that. Especially low performing leaders who use traditional "management" styles of micromanagement to push their people through change. When you push people through change, it is a poor experience. Highly effective leaders treat change as an experience. They lead from the front, buy-in to the change first, and then lead their people through that change.

Over your career, and your time leading a team, would you say you have developed yourself to become a better leader? Have you developed yourself in leadership and how to lead people (character)? Or have you developed yourself in management and how to manage people, processes, and things (competency)? Be honest with yourself when answering these questions. You are aware now that we can only *lead* people, we cannot *manage* people.

The reason that most people in leadership positions prefer to *manage* people rather than *lead* people is because, management is easy, and leadership is difficult. To lead others and to help them grow, you must first lead yourself, and work on yourself every day so that you grow. Be honest with yourself. Do you lead yourself, and work on yourself every day?

Have you accepted the responsibility to help develop everyone on your team in the area of leadership (character)? Do you take the easy way out and *manage* your people instead of *leading* them? Do you only develop them, so they are able to do the job (competency)?

If you want to strive towards becoming a highly effective leader, then you need to embrace a new way of thinking. You must help yourself and your team to experience a new kind of change. By embracing a new way of thinking, it will be your role to help your people to embrace a new way of thinking too. You must constantly be developing yours and your team's minds. As you work on developing minds, you are freeing up space in the mind to also develop new, or continuously improving old processes. Experience a dramatic change in your team and keep going.

From now on, if you want to implement a new idea, change, or process, you will need your team to buy-in to you and the idea, change, or process, and what the benefits are. To get your team's buy-in, you need to inspire them and motivate them. If you have been *managing* your team rather than *leading* them this is going to be a lot more difficult. You need to change your way of thinking first. You must lead yourself and motivate yourself before you can motivate your team.

We have discussed how difficult it is to get a team to buy-in to a new change. As you are aware, most teams will resist new change because they are not bought in, and do not want to change. So, make it as easy as possible for yourself. Only when you have bought-in to yourself and the new change, can you lead your team to do the same.

With me writing this book, my other books, and creating my other materials, I am trying my best to inspire you and motivate you to buy-in to me. As you are reading this book, it is my role to help you release the potential you must become a highly effective leader. I want to set you down the right path on your journey to becoming a highly effective leader. When you release this potential, you can then help your people to do the same as you and set them on their journey to becoming highly effective leaders.

At this moment in time, does your team follow you because they *want* to (leadership), or do they follow you because they *must* (management)? Again, be honest with yourself. If you want your team to buy-in to your idea, change, and your vision, then they must first buy-in to you. If they buy-in to you first, then they are following you because they want to, and you are on the right path to becoming their leader. If they do not buy-in to you, then you unfortunately are not their leader. You are just someone who is filling the leadership vacancy.

It took me several years while being in my first couple of leadership positions to figure out why some of the changes I wanted to implement, did not work…They did not buy-in to me first. I wasn't working on myself every day, so I wasn't leading myself first, and I wasn't really buying-in to the change first. It felt almost like bringing in change, just so I could say I brought in change.

After I had spoken with my mentor, and he helped me to realise that I needed to get the team to buy-in to me first, I knew exactly what to do. I needed to change the environment. I needed to create an environment that encouraged continuous improvement for ourselves, not just our processes. I needed to lead by example by working on myself, and then helping my team to work on themselves and develop. Only when I did that would the team buy-in to me, and eventually I could lead them through the change.

With this website and my books, I am creating an environment to help you continuously improve yourself. I am not creating an environment to help you continuously improve your processes. Continuously improving yourself is a part of your personal growth and is a personal growth principle.

When we focus on continuously improving our processes, we are expecting to improve our results. Now, we need to change how we think of continuously improving ourselves. We need to get comfortable and believe in the process that when we continuously improve ourselves, we can also expect to improve our results.

A highly effective leader's mission is to inspire their team and motivate them to continuously improve. Make it your mission to do the same.

Low performing leaders focus solely on the results. They do not focus on the people at all unless they want them to do a certain task. So, they will arrange for them to learn the new task but will not help their people to develop their character and leadership. Low performing leaders don't even focus on developing their own character and leadership either. All they are interested in is using their people to make themselves look good to the leaders of the organisation. I know this because I have been manipulated many times.

It is a horrible feeling when you know you are being manipulated. That's how you can tell it is a manipulation because it is a feeling that you have that something isn't right. But I didn't say or do anything about it because I didn't want to upset my boss, or "rock the boat" in any way. So, I just got on with it and let myself be manipulated. However, after a while I needed to snap out of it and think differently. I needed to believe in myself a lot more. I needed to lead myself better. I needed to work on myself so I wouldn't be manipulated again. I needed to become a highly effective leader and have respect for myself and my people.

Highly effective leaders lead their people through their strong character, and they have integrity. No matter what kind of environment they are working in, they truly stand up for what they believe in. They will never let themselves, or their team be manipulated. They will never manipulate anyone else either. Even if the senior leaders of the organisation are low performing leaders, a highly effective leader will not allow any kind of manipulation happen.

Most of the organisations I have worked with have had traditional leadership team setups. Their senior leaders focus on the results, not their people. The senior leaders do not have respect for their people because they don't develop their people. I tried my best to ensure that the team I led did have those opportunities to develop and did not follow the norm. I know this because I respected my people, and we worked on ourselves, and developed ourselves every day.

It is easy to tell if your leader respects you, or if they don't respect you. You can feel it. You can tell whether they are being genuine or being fake.

When I lead a team, I do my best to make sure that they *feel* that I respect them. I do not want them to feel that I am being fake. When I know that my team *feels* that I respect them, I can also *feel* that they respect me. When respect goes both ways, that is when I find it easy to help them develop themselves, and we can achieve great results together. Respect both ways is what highly effective leaders focus on and count on.

If we want to get things done, then we should focus on doing. Not focussing on talking about doing or planning to do. Just do it!

Respect your team to develop your team

How you treat your people is extremely important if you want your team to be successful. If you treat them with respect, and see the potential of who they could be, they will follow you. If you don't respect them and treat them on how they appear to you initially, they will not follow you.

There has probably been a time in your life when you have heard the truth from someone, and it hurt. It's probably hard to hear, but it hurt you because it was supposed to hurt you. I know the truth has hurt me on several occasions.

Do you feel that the organisation you work with respects you and all of your colleagues? Do you receive leadership development training, either in a one-off intensive course or an ongoing programme? If you don't, then your organisation does not respect you or your colleagues.

If your organisation respected you and your colleagues, then it would provide you, your colleagues, the middle management, the senior management, and the executive leadership team development training. Everybody in the whole organisation should be trained in leadership, personal growth, character development, communication, and productivity.

Organisations that have low engagement with their people, and a high turnover, do so because they do not develop them. This is the norm for most organisations throughout the world, which is unfortunate and must change. They must learn from organisations who do have respect for their people and provide them development training. Especially training in developing the character of all their people, not just the senior leaders.

When I was engineering technical manager for a UK train operator in London, I had a big team that were based in London, Essex, and Norwich. So, they were spread around the south and east of England. I had a team of managers who I led, and they led their technical teams. There was one team, the systems engineering team, who was led by a systems engineering manager, Andy. I had known Andy from my engineering graduate days, and we got along with each other fine. It wasn't until I became his leader that things started to change.

This was Andy's first job as a manager. Before he was systems engineering manager, he had been an engineer. He was chartered like me so was very experienced in the world of engineering. However, during his career, he had received very little leadership development (character). But had received lots of engineering, and technical development (competency).

Unfortunately, Andy wasn't doing as good a job as I thought he would. He was a great engineer, so naturally I and everyone else thought he would be a good leader. But this wasn't the case, and a lot of people were noticing, mostly me.

When I addressed this with Andy, he retaliated against me. He felt like I was singling him out, and brought up mistakes that other people were making, rather than focussing on himself. I didn't expect this retaliation from Andy, but it was something I had to deal with.

After we ended our initial conversation, I went away and thought long and hard on how I could help Andy, and how to resolve this. I spoke with my boss, and we decided that we would bring in a consulting company who could help Andy with his leadership skills and help him organise his team better. We wanted to see better results from him and his team, and we thought this help would be the answer.

So, when the company came in to meet the systems engineering team, I could feel that there was an atmosphere, and not a good one. The consultants presented to us what they were going to do to help, and everyone kept asking awkward questions and saying things like, *"We've seen this all before"* and *"This is money for old rope,"* and the old favourite *"This will never work."*

After the meeting ended, I took the consultants aside and we agreed that this was just the first day, things will get better as we go through the new processes and training. As you may be able to imagine, things did not get better. They got worse, much worse in fact.

Andy and the systems engineers were becoming very disruptive to the rest of the technical team. They wouldn't do work on time, sometimes they wouldn't do work at all. The atmosphere in the team was toxic, and it was having a big strain on me. I was getting criticism from my boss, his boss, the other engineering leaders, and it kept piling up. I couldn't figure out why Andy and the systems engineers were behaving like this. I thought I was doing a good thing by bringing in consultants to help them.

There was even an occasion when one of the consultants phoned me up to tell me that Andy had verbally abused him and insulted him. I couldn't believe what was happening. So, I confronted Andy with this accusation, and he denied everything. I didn't know what else to do. I let everything get on top of me and I even had to take time out with stress. It was during this time out that I learned why the team reacted with such retaliation to the consultants.

Do you know why?

You guessed it.

I hadn't involved the systems team in bringing in the consultants. I hadn't asked their opinions, their ideas on how we can improve, and I hadn't received their buy-in. I didn't show them or Andy the respect, and that I wanted to help them develop. I hadn't thought about character development, I only thought about competency development.

To be honest I didn't even buy-in to bringing in consultants myself. I just went with it when I had the discussion with my boss. I was so disappointed in myself, and really got down on myself. Andy was eventually moved out of the systems engineering manager role and was replaced. It was my boss who made that move.

This was a huge lesson for me. I knew from then on that I needed to help all my future teams develop their character and competency, not just competency. I knew I had to work on myself every day to develop my own character first, before I could help others develop their character.

Lack of communication, lack of engagement, blaming, resistance, and bad mouthing from your team is a sign that you as the leader are lacking in character. Your character is weak, and you must work on it to make it stronger. The starting point for this is to show respect to your team, and your team will in turn respect you back. By growing yourself first and developing your character first is how you become a highly effective leader. If you make character development of your team the norm, then your influence will increase around the organisation and other teams, departments and leaders will follow your example. Character development and respect for Andy and the systems engineering team is where I should have started.

Highly effective leaders focus their efforts on character development because they know this is the best way to increase their influence. The more they increase their influence with their team, the more successful they will be. The more successful they are, the more the influence of the leader and the team will increase into areas that they previously didn't have any.

Highly effective leaders know how important character development is, and so should you. From now on, focus 80% of your time on character development and 20% of your time on competency development. That is for yourself, and your people.

It's not about the results that you get, or the money you make for your organisation. It is about the lives you touch and the positive differences you can make in people that really counts.

Increase your influence with trust

If you want to build strong relationships and trust with your team, then you must develop yours and your team's character.

You cannot influence your team to become better people, more productive, be inspired, be motivated, or trust you if you have a weak character. Without the trust of your team, you will be a manipulator, not a motivator.

If you want amazing things to happen for you, your team, and your organisation, then your influence needs to be on the increase every single day. You need to keep up the influence momentum every day. Momentum is a leader's best friend. Stagnation is a leader's worst enemy. However, you cannot do this on your own. You will need the help of your team, and the support from your senior leaders too. So, to receive the support you need from your senior leaders, you must increase your influence with them.

When you have learned to increase your influence, it will then be your responsibility to teach every person on your team to do the same. By having a team who know how to increase their influence, will automatically increase your influence through them and with others throughout the organisation. Learning the correct leadership principles will enable you to increase your influence with anyone, including your senior leaders, even those at the very top of the organisation.

Highly effective leaders understand the criticality of increasing influence. Influencing and helping team members to buy-in to them, trust them, and follow them is so important to the success of the team and the organisation.

The important word I just mentioned…. trust. Trust is the foundation of leadership and influence. If your people do not trust you, they will not follow you because they want to. They will only follow you because they must, and that is not a good place to be in. As highly effective leaders, we need to accept the responsibility of the team. It is up to us to deliver the results that our senior leaders want. Without the team trusting each other, and especially the leader, then they are not going to produce the results. Low performing leaders will avoid the responsibility of the team.

Low performing leaders who avoid responsibility for their team and the team's results, create distrust with the team and the organisation. A low performing leader's influence decreases by the day when they are not trusted. Are you willing to accept the responsibility for your team, and the team's results?

What you are learning from this book will help you begin building new relationships with others, and to make your current relationships that you have with your team, organisation, friends, and family to be stronger.

If you want the results within your team to improve, and continuously improve, then you must gain the trust from your team as soon as possible.

When building relationships with others outside of your team, treat it as a great opportunity to plan. You may want to ask one of your colleagues in a different team to join your team. So, when you have already done the work by building a relationship with this person, you will already have gained their trust when they start with you. This will make the process a lot easier and smoother.

When I was engineering technical manager, it took me a good while to start building trust with the team because I didn't make it a priority right from the start. I should have gone to every single person within the big team that we had and built a relationship with them all. Not just the select few. That was a mistake. Don't make the same mistake as I did. Build and relationships and create trust with your team as soon as possible. Make it a priority of yours.

When trying to build trust as soon as you can with your team, look at it as a two-way street. The more work you put in to building relationships and building trust, the stronger your relationships will become, and the more trust you will receive. You must give to receive. You cannot expect to receive without giving.

Much like when you go the gym. You cannot get into shape if you don't work out and put the effort in. If you don't put the effort in to build trust, you will not be trusted.

When you put the effort in to build strong relationships, and build trust, your influence will automatically increase. It is the same process with every single person you meet or encounter. Including your senior leaders. You must first make the effort with your senior leaders to build trust. Don't expect or wait for them. The longer you leave it, the harder it will be to build trust.

As you are increasing your influence and building trust, you will come across people who want to push you down. This happens in all walks of life. They may say something to one of your team members or colleagues that will automatically create distrust. However, when this happens you will need to build on that relationship again until you erase the distrust and create trust again. I have come across these people, and it is just out of jealousy that they do this. So, take it as a compliment.

You will also come across people in your teams, or in the organisation who have been encouraged to be resistant in their careers. They resist the senior leaders, they resist change, and when you become their leader, they will resist you. When I have worked with these people, it is much, much harder to build a strong relationship and trust with them. However, you must not give up. It will be a challenge and an obstacle, and you must overcome that obstacle to keep your influence increasing.

Other leaders within the organisation may say things about you to your team that may cause distrust, because they are weak and resist anything new to happen within their organisation. They resist change, and they believe that talking you down will stop any positive change from happening. This has happened to me a few times. So, before I even started, I was already distrusted.

Again, you will need to work a lot harder to build relationships and trust when this happens. It may sound unfair, but it's reality. You must take the higher ground and rise above those weak leaders by becoming a highly effective leader. You must become the leader that you want to see and who your teams want to follow.

I found it extremely hard work to turn distrusting me into trusting me. But I didn't give up and I was able to do it with the people I put the most effort into. If you want the team to produce successful results, then building trust must be the priority, and you must do it as soon as you can.

Whenever you start leading a new team, they will initially have a skeptical mindset about you. That is just natural, you probably had the same mindset when you met your new leaders. They want to find out if you respect them, if you can help them, and most of call if they can trust you.

It will be up to you to change their mindset and prove to them that you do respect them, you can help them, and that they can trust you. How are you going to give them that proof?

If you want to implement a change, you must be trustworthy. You cannot change anything without having the trust of your people.

Your team must visualise your vision in the same way as you. They must feel the same things you feel

CHAPTER 5
COMMUNICATION IS NOT ENOUGH

Be very clear of your intentions

When a leader provides his/her team with a compelling vision, goal, and purpose, their team will become followers.

A highly effective leader will live the vision, goal, and purpose every day from the front and set the example. Once the example is set, the followers will follow in the same way through the leader's inspiration and motivation. This is what aligns the leader and the team.

When I was told that I would be moving into my first leadership position, I was so excited. I felt that I was on the next step of climbing the ladder. I was excited to see what difference I could make in engineering. I was excited to make our trains better. I was excited to impress the senior leaders. It was all me, me, me, and what I could do to make things better.

I didn't think about what I could do to help others become better. I didn't understand that my new role was not about engineering anymore. It was about human beings. It was about the people I worked with, and the people who would be on my team.

On my first day in my new leadership position as an engineering production manager, I very quickly realised that most of the other people I worked with were not as excited as I was. They didn't share my optimism to make things better. They didn't like everyone on the team. The team hadn't met me before, and already they didn't like me. They didn't like the other engineering leaders, and they certainly didn't like the senior leaders. Some members of the team didn't even like the place where they worked (this was when I was in Edinburgh, Scotland).

It was a rude awakening, and my optimism and enthusiasm very quickly dropped.

On that first day as an engineering production manager, I attended the morning meeting. This was led by the production manager on shift, and he would assign the work to the team for the day. As it was my first day, I was shadowing the production manager to learn *"the ropes"* and get familiar with how the engineering depot worked. Anyway, the meeting started and before anything was said, one of the team members put his hand up to ask a question, his name was Tam. His question was *"Who the F*** is he and what is he doing here?"* He was pointing at me. When he said that, I froze. I was so nervous, trembling and confused, I didn't know what to say. I hadn't been confronted like that since I was 16 during my apprenticeship.

The production manager introduced me as the new engineering production manager on shift. Then Tam asked me, *"Have you been a manager before?"* I said, *"I have but I haven't had my own team before."* His reply was, *"What makes you think you can come here from Liverpool and manage us then?"* This wasn't going very well, and I had to do something.

I asked Tam to come with me for a quiet chat. When we were alone, I said to him *"I know what you're trying to do. You're trying to embarrass me in front of your team because I'm the new manager. Well, it won't work with me. So, we can both start again on the right foot, and go to work, or we can have a much deeper conversation about this, and I can escalate to the senior leaders your behaviour.... it's your choice."* Fortunately, he decided to start again with me, and we both made a clean slate of it.

I must admit, I could have probably handled the situation a lot better. But it was this confrontation that really drove me to learn and study more about leadership.

After a few weeks of being in this new role, I learned that the production teams didn't really trust us as production managers. What was dis-heartening was that the other production managers knew it but didn't really do anything to address it. I wanted the team to trust me, and I wanted the team to like me. But I knew I needed to be the one to build the trust with the team and not the other way around.

It was difficult because I started to build a relationship and trust with two members of the team, but as this was happening, I was creating distrust with

A LIMITLESS MINDSET

other members of the team. Trust is a "two-way street", and we only trust the people who are like us and share the same values. We all have different values. However, the one value that everybody shares is us.

So, how do you build trust with everyone on your team, who all share different values? You need to be very clear of your intentions with the team. You are the leader of the team, but you are also on the team's level, and are one of them. Any successes that happen from your team, it will be the team that is responsible, not you. Your main intention and objective are to help everyone on the team to become better people, and also a better team.

Being intentional and deliberate with your team like this, is how the team will understand exactly what your intentions are. Be very clear on **why** you are doing what you are doing or saying what you are saying. When you tell them the reason for your action, you are confirming what the team were already thinking. Don't just do or say something without a reason. Otherwise, the team will not understand why you are doing what you are doing, and will make assumptions about you, what you say, what you do, and why you're their leader. Most of the time these assumptions are negative.

The only way to avoid these negative thoughts by your team is to **be very clear of your intentions**. Always tell the truth with your team and be open with them. What we as leaders need to come to terms with is, whenever we tell our teams our intentions, they don't believe us. What they believe is their own interpretation of what we say. Most of the time it is the wrong interpretation. When you are reading this book, it is likely that you don't believe what I am saying either. This is a human fact, and something we all need to come to terms with. We do this without knowing we are doing it.

Do you agree with what I just said? If you do, then you believe what I just told you. If you disagree, then you don't. You need to realise that what you have just told yourself you believe, wasn't what I really said. It was your interpretation of what I just said. This is the same when you talk to your team. They interpret what you said, and most of the time it is wrong.

To overcome this, you need to **do** exactly what you **say** you are going to do. The action you take needs to be right down the line of what you say. When you align your action exactly to what you say, this is what builds trust with your team. Your team will then interpret what you initially told them in the right way. That is why we have phrases like *"walk the talk."* It means, do exactly what you say you are going to do. This will increase your influence as you build the trust up.

If you say one thing and then do another, then you are creating distrust, and your influence will decrease.

There are two ways you can be intentional. You can either be intentional to motivate your team, or you can be intentional to manipulate your team. When you manipulate your team, it becomes an *"us and them"*, toxic environment and culture within the team. When you motivate your team, it becomes an *"us and us"* or, *"we"*, inspirational environment, and culture.

When you are very clear of your intentions, you are providing your team with your compelling vision. They know what you want to achieve for the team, and why you want to achieve it. When you respect your team, they will trust you. When you trust your team, you value them. When you value your team, they will respect you.

Once the leader has been very clear of his/her intentions and what the vision is, the team will then take that vision as their own. They will live the vision and protect it. The leader will do the same.

How do you inspire your team?

You do not always have to be on the higher ground to be great.

When a highly effective leader takes over a new team, the first thing he/she wants to know is, who the team are. They are not interested in what they do initially. They want to get inside the minds of every single individual on the team and learn about them as a person. The first meeting they have with the team is not about him/herself, it is about the new team he/she has just taken over. The highly effective leader will ask the team a lot of relationship type questions. By doing this, they and the rest of the team can learn things about each other that they didn't already know. The team are also invited to ask questions of the leader, so that they can learn about him/her in the same way. This is all about humility, which is one of the great traits of a highly effective leader. It is not about results, or the job.

Humility is not a trait of a low performing leader. If you ask a low performing leader what humility is, they might struggle to give you the correct answer. Especially the low performing leaders I've worked with in the past. In my experience, when a new low performing leader took over a team that I was part of (and there have been quite a few), their initial meeting with us was all about them. One leader gave us a presentation that was all about himself, and how he got to where he was. He didn't ask us about who we were or our careers. My first impression of a low performing leader is, they are very arrogant, they have a huge ego, and they have a lot of pride in themselves. That then leaves no room to have pride in others, especially their team. We were treated as numbers, instead of people. It is very uncomfortable to be around them. Whenever they are asked a difficult question, they will avoid it like the plague. Their communication skills are very low, and they are not clear of their intentions (as we discussed previously).

Highly effective leaders ooze confidence, and low performing leader's reek of arrogance. The difference between arrogance and confidence is very tiny. Highly effective leaders know these differences well, and they know how to stay on the right side of confidence. Low performing leaders don't know these differences, hence they come across as arrogant.

When low performing leaders try to communicate with their team, what they are doing is not communicating. They are telling. Communication is a two-way street. Telling is a one-way street, very much like dictatorship. Highly effective leaders are excellent communicators, but they can also go beyond communication and connect with their people. They connect deliberately because they know that connection doubles communication.

To motivate our people, we need to communicate with them. Motivation is external, and when we motivate people we are talking to the person's thoughts, or their conscious awareness. It is our words that they listen to that could change their behaviour is some way. Communication and motivation can only be done externally to the person or people.

To inspire our people, we need to connect with them. Inspiration is internal, and when we inspire people we are talking to the person's feelings, or their sub-conscious awareness. How we make our people feel from what we do, not just what we say. We can inspire people from the tone of our voices, our body language, but most of all from our actions. Connection and inspiration can only be done internally to the person or people.

Everyone throughout the world can communicate, even low performing leaders if they choose to. When we communicate however, it doesn't necessarily mean we are building trust or building a relationship with a person. There are different types of communication. But, to communicate all we must do is share the information we have with another person. When we do that, we have officially communicated.

Not a lot of people throughout the world can connect with others. Highly effective leaders know how to connect with their people because they learn how to do it beforehand. They study connecting with people, then they practice connecting with people until they can do it effortlessly. Connecting with people is how you make a person feel good, or positive. It is about feelings. When we deliberately connect with a person, we are building a relationship with them. What do we build when we build a relationship? Trust. A highly effective leader will connect with their people every day.

I have worked with a lot of low performing leaders who *"tell"* us what to do. Unfortunately, I have only worked with a few highly effective leaders who *"inspired"* us to do the right thing. I want to create more highly effective leaders who can inspire more people, and hopefully one day outweigh the low performing leaders.

When a highly effective leader inspires their team, it is very easy for them to get the team's buy-in to new changes or new ideas. They trust their leader, and they follow him/her because they want to. Low performing leaders find it very difficult to get their team's buy-in because they don't trust their leader, and they don't follow him/her. When this happens, the team become impossible to work with. I have experienced this firsthand on many occasions.

A lot of low performing leaders are in the leadership role because they were forced into it, or nobody else could take up the role. This too is unfortunate because you have a person who is supposed to lead and inspire a team, when they don't want to. That is a recipe for disaster. Nobody should ever be forced into leading a team. It will never work.

I speak a lot about having respect for your people. When forcing one of your people into leading another team, when they don't want to be not showing them respect. Not giving them a choice is manipulation and can come across as a threat. You MUST find a person who wants to lead a team, who wants to motivate people, who wants to inspire people. Someone who wants to help others and help you too.

A highly effective leader will always show respect to their people. They do not want their people to ever think that he/she doesn't respect them. So, to avoid this they will never force one of their people into doing something that they don't want to do. Especially leading a team. Leading a team is a huge commitment and is very difficult. It is something you must want to do and must never be something you have to do. Much like following a leader. Following a leader must be something you want to do, not something you have to do.

When I wanted to take over my first team, and I thought I was ready, I was living in Liverpool. It was 2011, and I went to my engineering director Kevin and discussed it with him. I had a strong relationship with Kevin, and he had helped me to develop and didn't force me into taking up any kind of role that I wasn't ready for. He had helped me build up my confidence, and he inspired me. In other words, he connected with me. He went beyond communication and connected with me, as he did with many others, and still does to this day.

Forcing people to do things is not how you inspire people. Connecting with people is how you inspire them. When connecting with a person, you must make it about them. You must listen to them well. Let them do most of the talking, and you do most of the listening. When you listen to the person, they feel cared for, when they feel cared for, they feel inspired, when you inspire them your influence will increase.

Connecting with a person is a great opportunity for you to learn from them, and about them. How do you do that? You listen to them. Interrupting them, or deliberately cutting them off will break that connection. Control yourself and you will connect with your people.

How do you keep your team motivated?

Listening to our people's frustrations and helping them overcome their obstacles is how we motivate our people.

To help our people become the best people they can be, we must help them to remove the obstacles in their way. We must help them to solve their problems. However, we must not remove the obstacles and solve the problems for them. Our role as a highly effective leader is to set our people in the right direction and assist them on their journey. We must help them to help themselves come to solutions that will overcome their problems, obstacles, and frustrations.

One thing I have learned over the years of being a leader of teams is, the people in my teams have had a lot of frustration. I am sure you have experienced the same, with the teams you have been part of or have led. I have had a lot of frustration over the years when in work, and it is not always built up because of work. A lot of my frustration has come from things that have happened in my personal life. For example, breaking up with girlfriends or family issues.

But what I couldn't do was to talk about my frustration to a leader who was willing to listen. For years, as a leader I didn't encourage my team to talk about their frustration either. Not because I didn't want them to, the reason was I just didn't think about it. I didn't think that it was my responsibility to listen to my team about their frustrations they were feeling from their personal lives.

Now, I believe it is my responsibility as a leader to listen to my team about anything they want to talk to me about. If they need me to listen to them so they can get things off their chest, then I will be there. If their frustration is with work issues, personal issues, or any other issues, I wanted to show them that I was there for them. Not just *tell them*. I wanted to *show them*.

When I changed my mindset, and I started to *show my team* that I wanted to listen to them, and help them with their frustrations, I found it easier to get my team's buy-in. Every person has problems to solve, obstacles to remove, and frustrations to vent. This was my way of showing my team that I cared about their problems, obstacles, and frustrations. I wanted them to know that they didn't have to try to solve them on their own. I was there for them. I started to enjoy listening to my team's frustrations because I knew we were going to get to a solution together. The team got excited when they talked of their frustrations with me because they knew we were going to get to a solution too.

If you want to increase your influence with your team and get their buy-in on any new changes or ideas, stop talking about work all the time. You need to become more personal with your team. Your relationships at work cannot stop at work. To build stronger relationships, and stronger trust, and increase your influence, listen to your people.

By listening to your people, and allowing them to vent their frustrations, you are automatically motivating them. Following your conversation, your team member will be motivated to go back to work and do a great job for you and for the team. When the team are motivated, they are bought in. When your team are bought-in, it will be a lot easier to leverage their abilities and skills to improve results and help the organisation.

If you read this section and then carry on working with your team in the same way, and not listening to their frustrations in their life, then the best you will do is communicate with them. If you change your ways, and allow the team to talk to you, and you listen to their frustrations then you are connecting with them. It is very uncommon in the corporate world for a leader to listen to their team's frustrations. However, you are not a common leader, you are an uncommon highly effective leader.

When having a conversation with one of our team members and they are talking about their frustrations, we must be very careful to let them do 80% of the talking, and we do 80% of the listening. That's the difference between a highly effective leader and a low performing leader when it comes to conversations like this. A highly effective leader opens his/her ears and listens. A low performing leader opens his/her mouth and talks.

When we listen 80% of the time, we are *showing* our people that we care about them. We are *showing* them that we want to help them.

I wish I would have known from the very beginning of my leadership career that listening to my team would have more of an impact than talking to them. I wish I would have known that giving my team the opportunity, to talk to me about their frustrations, would motivate them more than anything I would ever say to them. But now, I am happy that I do know these things, and that I can teach you and many others that listening is far more powerful than talking.

When it is our turn to talk 20% of the time in our conversations, we must be very careful in what we say because what we say really matters to our team. Especially when it comes to a person's frustrations and problems.

What we say can be the difference between getting our team's buy-in, creating trust, and increasing our influence. Or losing our team's buy-in, creating distrust, and decreasing our influence.

To provide you with an illustration of what I mean, I want to present you with two scenarios. One of them really happened with me when I was engineering production manager in Scotland. The other scenario is what really should have happened but didn't.

Scenario 1. It is my daily morning brief with the engineering production team at 8:30am. I stand in front them with my presentation behind me. I say, *"Good morning. Erm…come on guys, I have said good morning to you, why aren't you saying it back?* (Silence) *Today's work is on the board, work it between yourselves who is doing what. I will be around to your work area at 1pm to see how you're getting on. Let's get to work guys."* I walk out of the room back to my office.

Scenario 2. It is my daily morning brief with the engineering production team at 8:30am. I stand in front of them, and I hand out a copy of my presentation for everyone to see and read. I say *"Good morning guys, I want to thank you again for your contributions last week. You all put in a great effort again, and the senior leaders noticed. I have encouraged them to come to the depot to come and see us, so we can show them what we're doing. How was your weekend?* John replies, *"It was great tom, I took my son to see the football.* I reply, *"Great, glad you had a nice time. I have printed out the workload for today and assigned each member of the team to the tasks that you have strength in. If you need my help and assistance, please don't hesitate to ask. I will be walking around to see how everyone is, so if you need to talk to me about anything,* *just grab me. If you want to share your frustrations with me, I would love to listen and help you. I'm here for you, so let's all have another great day."*

What scenario would you rather listen to if that was your leader addressing the team at 8:30am? What scenario would your team rather listen to you say to them at 8:30am? What scenario would inspire you more? What scenario would you have a better chance of getting your team's buy-in? What scenario would you trust your leader more from? What scenario would you find it easier to contribute to? The answer is easy, Scenario 2.

What we say to our team, and how we share our vision to our people is extremely important when we want to increase our influence. As a highly effective leader, we have followers. Our followers will put their hand up and volunteer to contribute to our vision and do their best to make it a success. Our followers will already be bought-in and motivated because we have listened to them and said the right things. They make our vision their own and make it their mission to achieve the vision.

There is one thing having a vision. It is another thing having a vision with a plan. It is the only thing when we have a vision, a plan, and we consistently act on that plan to bring the vision to life.

Talking more than you listen demotivates. Listening more than you talk motivates

CHAPTER 6
INSPIRING YOUR TEAM TO BUY-IN

Listen more than you talk

To connect with your team, you must listen 80% of the time.

Early in my leadership career, I thought that it was the leader's role to do most of the talking when in conversation with my team, or as individuals. I thought it was the leader's role to stamp their authority on the team, and that the team should listen to the leader. This was a big mistake on my part because we now know that the leader's role is to do the exact opposite.

I didn't know what I didn't know back then, but I was continually learning, and I learned quite fast that it wasn't up to me to do most of the talking (20% of the talking). It was up to me to do most of the listening (80% of the listening). I thought of myself as the "expert" because I was in the leadership position. But I was by no means the "expert". I had a team full of experts, so I should have listened to my experts a lot more and leveraged their expertise.

When I started reading leadership books and taking my leadership development more seriously, my listening skills started to improve. I was also improving at letting my team members do 80% of the talking in our conversations. Especially when we were talking about their frustrations, or personal issues. My leadership styles were improving every day, as I worked on myself every day.

Asking questions of my team, rather than giving them the answers or suggestions was something I developed too. Especially open questions that would enable my team member to draw out more information from inside themselves, and eventually work out the solutions themselves with my guidance. By doing this I was respecting my people, and they would respect me because I listened to them.

When asking your team members questions and helping them draw out their own answers and conclusions, it takes more time than just telling them the answers. But it is worth taking more time because you are going beyond communicating when you listen 80% of the time. You are connecting with your team. When connecting you are increasing your influence and building up stronger trust with them.

What you will find is, when you make stronger connections with your team, they will become more successful. They will achieve more, so again it is worth taking more time with them by asking them open questions (what? who? where? why? how?). When the team start to become more successful and achieving more than they thought, you will find it even easier to get them to buy-in to you, your ideas, and any changes you want to implement.

You and the team will grow together, and you will grow closer. You will no longer need to set the direction for the team, they will happily follow you on their own down the right direction. The team will choose to stand beside you and unite as one team. When that happens, you will know that your influence is increasing with the team every day.

To return the team's loyalty, trust, and faith in me as the leader, I would do my best to help the team as much as I could. I would help them to solve their problems. I would support them and use my leverage to promote any new ideas that the team had. I would ensure that my presence wasn't an obstacle to them at any time. I would remove obstacles for the team when they needed me to. I would constantly be available to listen to the team. I would continue to ask the team open questions when it was necessary.

There were occasions when I would let team members lead the team, and myself when it was needed. When I listened to the team I was being led. By listening to them team, I was creating leaders. By writing this book for you and building this website, I want to create a highly effective leader out of you.

Since working in the engineering sections of the rail industry, one thing I have always been associated with is train performance. Have trains performed well in service? Were there any delays? Were there any failures/breakdowns? Were there any cancellations? What has been the worst performing train this month? What component has caused the most failures this month? What are we doing to improve performance?

These are all questions that I heard every day as part of the rail industry. When I was engineering technical manager working in London, there was one fleet of trains that had been performing very badly for about three months straight in 2016. The leaders of the business wanted to know why. So, they arranged a meeting for my managers and I to present to them the reasons for poor performance, and what we were doing about it.

There were so many reasons for the fleet's poor performance; doors failing, couplers not interlocking properly, brakes issues, wheelsets wearing, etc. This fleet had been poor for years, but the engineering team had just done enough to keep it going. They didn't go the extra mile to work as a team to improve their own, and the train's performance beyond expectations.

I knew I had a challenge on my hands. So, instead of laying into the team and demanding ideas of them, me doing most of the talking, I decided I would change it around. I decided I would let them tell me, and I would listen to what they thought. I wanted to find out from them why they hadn't gone beyond expectations. I wanted to know their ideas so we could implement them and turn performance around.

The team felt like they had never been listened to before, so it was difficult to get them to open up at first. This was new territory to them, they didn't know whether I was trying to manipulate them, or if I was being genuine. So, I stopped. I decided that it was up to them to come back to me with a solution on how to improve performance, and they were going to present to the senior leaders.

When I made this decision, the team were not happy at all. They didn't want to work on this as a team, and they didn't want to present to senior leaders.

So, they got together in a meeting room to brainstorm their ideas, but all they did was bad mouth me for making them do this. They couldn't come out with any ideas, so they wasted the whole time they were in the room together supposedly brainstorming.

So, when I went in to their "brainstorming session" and found that they were just moaning, instead of working, I thought I would try again to get them to open-up. I asked them open questions, and then would listen to them.

For about three hours I listened to the team tell me why they shouldn't be doing this, why the trains performance will never improve, why the fleet of trains should be scrapped. Every negativity that they could come up, I listened to for three hours.

So, I made another decision. I decided that I agreed with what they were saying. I said to them, *"Ok guys, if that's what you're telling me, then we can't go any further. Let's end the session here."* They were surprised at my reaction because they were expecting me to fly off the handle. But I didn't. I then told them, *"Make sure that when you present to the senior leaders that you tell them there is nothing we can do."* They looked at me and said, *"We can't say that to the bosses, they will fire us."* I said, *"You're right. But you have decided there is nothing you can do, and I don't lie to my bosses. So, make sure you tell them the truth."*

Straight away, they decided they will start the brainstorming session again and come up with a solution. So, I left them to it. The next day, they had a full presentation of the reasons for poor performance, and three new modification ideas that would help the trains to improve their performance. When they implemented the new modifications later in the year, the train's performance improved. All by working together as a team.

Leadership is about people, not numbers. Taking care of the numbers will not take care of the people. Taking care of the people will take care of the numbers.

Preparation is essential

Do you have a win-win mindset with your team? If you ask them to do something, is it for the good of you, for the good of your team, or is it for the good of both you and the team?

When you arrange meetings with people, either your team, other colleagues, or the senior leaders, how do you prepare? For years when I was in a leadership position, I would mostly just prepare on my own. Or I would discuss a certain issue with a team member, then I would go and have the meeting. But, when I was doing that, I was never, ever satisfied with how my meetings went. There was always something that happened that I felt made the meeting unsuccessful, and I couldn't put my finger on what it was. The result of these unsuccessful meetings was the actions taken from the meeting were either not completed or were completed to not a very good level.

Then, one day, I was running late (because I had not prepared well), and I turned up to *my meeting* totally unprepared. As you can imagine, the meeting did not go well at all. However, I could finally put my finger on why my meetings were not going very well....**ME**.

I was the reason. My preparation was nowhere near good enough. So, I decided that I would start to have a meeting before the actual meeting. People call it a pre-meet, but to me that is just a catch up, not a meeting. I needed to have an actual meeting before my actual meeting. So, I called mine a *pre-*

meeting, not a pre-meet. This would take up a lot more time and effort on mine, and the team's behalf but it was worth it. When I started to do this, others followed in my example too, and our meetings were far more successful.

The *pre-meeting* was especially important if I was trying to gain buy-in from my team or from others during the actual meeting. Having the pre-meeting allowed me to prepare how I was going to get buy-in, and how I was going to increase my influence. So, the influence and buy-in seeds were planted in the pre-meeting, and I could start to gain momentum then, instead of during the actual meeting which would be too late. I needed to build on my momentum from my pre-meeting, not try to begin the momentum during the actual meeting.

This principle is something I want you to consider doing because it can make you actual meetings become more successful.

When having your pre-meetings, always think of a win-win situation for all. The pre-meeting, or indeed the actual meeting is not only so you can have a successful meeting. It is for everyone involved to have a successful meeting. If you have this mindset during your pre-meeting, then the actual meeting will be a lot easier for you to gain buy-in and increase your influence.

If you do not have your pre-meetings, and do not have a win-win mindset, then your meetings will be just as difficult. They will become a lot more difficult too as you progress in your career.

Having the pre-meeting and keeping a win-win mindset is another way for you to connect with your people. You are going beyond communicating to connecting because you are willing to put the extra time and effort in with your team to prepare for the actual meeting. The team will feel this effort from you, and your relationships with each team member will grow stronger. Finding the time to have your pre-meetings is something you will have to plan for and make time for. Do not say to yourself, *"I don't have the time."* Or *"Where am I going to find the time?"* This will create a negative mindset for yourself, and you will put the pre-meetings off, which will decrease your influence, ruin your preparation, and buy-in will become a lot more difficult.

Most leaders (in fact most people) from all over the world invite people to their meetings from an Email request that goes straight into their diary. A lot of the time, people don't understand what the meeting is about, or why they have been invited. But the meeting request is from their boss, so they must accept and turn up. What happens then is, the meeting begins, nobody has a clue why they're there, or what to contribute. The leader does most of the talking, assigns actions to people, and then everyone walks out. However, when people walk out, they feel uninspired, they feel de-motivated, and most of all they feel disrespected.

I understand that leaders are busy, and that everyone else is busy too. But as I said earlier, pre-meetings, and meetings need to be planned for. We need to find the time for these meetings to make them successful. When we do this, we are showing our people that we respect them. When we don't, we are not showing respect because we are not willing to put in that extra time and effort to help them.

An example of this was when I started to organise my first team meetings. I thought it would be good to have a team meeting so we could discuss issues and see if we could help each other. So, the intention was right, but how I went about it was wrong. Very wrong.

I told my team that we would be having a team meeting…by email. So, nobody replied to ask why, or what we would be talking about because I hadn't spoken to them face to face about it. My email was very vague and gave just a few details. I didn't have an agenda, or any bullet points of topics to discuss. I thought it would be good for all of us to open-up. However, as you well now know, if you want a team member to open-up, you must build trust, build a relationship, and connect. You cannot do it in a meeting with the whole team.

Anyway, the meeting time came around and I was totally unprepared. How I could have thought I could make this meeting successful without an agenda, topics, or even telling the team a reason is beyond me now.

So, I began the meeting, and I told them why I wanted to call it. I said, *"I thought this would be a good chance for us to get to know each other better."* Everybody looked at me blankly and gave me the impression that they didn't want to be there at all. Some even laughed at me.

One of the team members said, *"Tom, I understand that you want us to get to know each other better, but you could have done it a lot better than this couldn't you? You haven't spoken to us, you haven't told us why you wanted us here, and you haven't given us anything really to talk about."*

He was right. I was basically "winging it" and the team felt that. They felt disrespected. I didn't have a pre-meeting with them, and I didn't connect with them beforehand. I didn't have a win-win mindset for all. So, I called the meeting off. This was the meeting when I put my finger on why my meetings were unsuccessful.

I turned it around following this meeting because of my pre-meetings.

Highly effective leaders understand the importance of putting in extra effort and investing more time with their people to have a pre-meeting, and then an actual meeting. They know that this is the only way to make it easier to get buy-in, build trust, and increase their influence.

They do not send out email meeting requests and then don't tell anyone what the meeting is about, or why they are invited. They know that by doing that, it is massively putting themselves in danger of being un-prepared. They know that by doing that, they are creating distrust. They know that by doing that, they are decreasing their influence, and their ability to get buy-n. But, most of all, they know that by doing that, they are disrespecting their people.

Highly effective leaders make it a priority to have a pre-meeting. They deliberately want to keep building on their already strong relationships, and make it even easier for themselves to get buy-in. During the pre-meeting they will set the vision, ask the team open questions, and most of all they will listen 80% of the time.

The pre-meeting adds to the culture and environment of the team and the organisation. It gives people a warm feeling that their leaders are willing to put the extra time and effort in to help them and the organisation to move forward.

How clear is your vision? Can others see it like you can? Being clear on your vision is a huge part of being a highly effective leader.

It's all about the people

Managers bring in change to processes and things. Highly effective leaders influence how you behave and think about processes and things

Do you understand the difference between managing and leading? The behaviours of a manager, and the behaviours of a leader are very different. However, as a highly effective leader, you must learn the ability to behave like a manager and behave like a leader at the right times for the right reasons. If you can do this, then you and your team can achieve great results. Your team will also be inspired by you to keep going and keep achieving these great results. This all comes down to the action you take, and what you say to your people.

If you don't yet understand the differences between leadership and management, then don't worry, you will. You will also be able to help others see the differences too by using real life examples, just like I am going to do.

When I took the role as performance manager at Siemens in 2015, it was my first leadership role when the team reported to me, and me only. This was following my role as engineering production manager, where I was part of a five man shift pattern of other engineering production managers. Anyway, after the first few weeks, and learning lots about the trains we were working on, and how they performed, I came up with a few ideas…all on my own. I hadn't discussed these ideas with my team, or with any of the other teams. My thinking at the time was, I am the boss, I need to come up with the ideas to improve train performance.

So, I go to my newish team and *tell* them my ideas, and that I want to implement these ideas as quick as possible. It was just a one-way conversation, me talking to the team. I didn't encourage or allow for any discussion (not on purpose), I just talked, and talked.

Next, after *telling* my team the idea, I then *told* them what I wanted to do to implement my new ideas. I directed exactly what to do and when, and when I expected them to have their tasks completed. Following completion of the tasks, I wanted them to report back to me to assure me that things were done.

When I think about it now, it frustrates me that I allowed myself to behave in such a way, and I was completely blind to it. I was so engulfed in my own "brilliant" ideas that I wouldn't allow anybody else to contribute. The fact that I was doing this sub-consciously makes it worse.

You can imagine how the team felt. They started to distrust me, probably talked about me in a negative way to others. Anyway, let's just say we got off on the wrong foot, and it was all because of me. I am happy to admit that. The ideas that I came up with didn't work anyway, so it showed me that I cannot do things on my own. I bet if I had involved the team from the beginning, the ideas would have been ten times better, and would have worked.

The example I have just described was not me leading my team. What I was doing was managing my team, and treating my people like they were things, processes, and even worse…numbers. You cannot manage people; you can only lead people. However, most *"leaders"* throughout the world manage their people, and do not lead them. Telling people what to do, by when, and to report back is micro-management. Micro Managers are very insecure, and weak leaders. They just don't know that that is what they are. Micro managers have huge ego's (as I had developed) and behave in this way because of their insecurities and weaknesses.

The way I should have behaved with my newish team was like a leader. I should have brought my team together, and *showed* them my idea, and then welcomed their opinions, suggestions, recommendations, criticism, and any feedback they wanted to give me. I should have listened a lot more than I talked. I should have put them before myself. I should have known that leadership is all about my people, not all about me. We should have made the decision on what ideas to implement together, and then worked out between ourselves who did what. Not me *telling* what to do and by when.

Highly effective leaders happily share their knowledge and ideas. They want their ideas to work, and they know that they cannot make their ideas work on their own. They need a team of people that they care about, trust, and have a strong relationship with. They need to have these things so that their team will buy-in to them and their ideas. Highly effective leaders involve their people right from the start, not at the end, or in the middle. That is how a highly effective leader makes things happen.

Highly effective leaders influence their people to think differently. They influence their people to think a lot deeper than they normally would so that they can think for themselves and come up with their own ideas and solutions. When this happens, they are opening the door for their people to take ownership and accept responsibility for their own actions and ideas. Allowing your team to provide solutions will increase your influence.

Influence is the foundation of leadership, and when you influence your team, they will take action on your influence. They do this because they feel respected by you, and in turn they respect you back. Leading and not managing is the only way you can demonstrate respect for your people.

When we are managing, we are improving our competency. When we are leading, we are improving our character. When we manage, we manage things and processes. When we lead, we are increasing our influence with our people. We cannot increase our influence with things and processes because they have no feelings. We cannot manage people because they do have feelings.

Even through our people are being paid for the job they do, we must treat them like they have volunteered, and are doing this job out of the goodness of their heart. That is how we want them to feel. We want our people to feel at the end of the day, and think to themselves, *"I would have done that job today even if I didn't get paid for it."* If we can get our teams to think like that, then that is true leadership.

Leadership is about people first, and processes second. It can never be the other way around. If it is, then that is management, not leadership.

When it comes to leadership, you must do what you say you are going to do. You cannot just talk the talk; you must walk the walk. The best way to do that is to lead by example, lead from the front and show your people the way. If you do not, then you are not leading. Would you rather lead your people, or would you rather manage your people? Your choice.

You want to make your leadership so good, that nobody even knows you are leading. Lead so well that your people feel that being a great team is the norm

*Share your dream with your people.
Show them how passionate you are*

CHAPTER 7
IT'S A TEAM EFFORT

Set your vision for the team

Passion is what comes across when you set your vision.

If you don't show passion when you are casting your vision to your people, they won't believe you. If you show them that the vision you are describing is your dream, then they will believe you and follow you.

The best leaders in the world (highly effective leaders) know that leadership is about helping others reach their full potential. They also know that is also about them reaching their own potential too. Do you know how to help your team, and help yourself to reach your full potential?

Part of a highly effective leader's purpose is to help their people reach their full potential and achieve amazing results because of it. A highly effective leader's passion is to continuously learn new leadership principles, apply these principles and act on them, and then teach these principles to their people so that they can do the same. By doing this they are increasing their influence and building trust.

Following my micro-managing of my team in Siemens that we discussed in the previous chapter, I wanted to ensure that I stayed on top of my leadership learning. I had slipped with my team, and I had let myself go into management mode of people instead of leadership mode. So, I started reading leadership books every day. I still do today, even if it's just for 30 minutes. This is part of how I work on myself. As you are now by reading my books. There is no end to learning about leadership which is what I love about it. That's why it is a lifelong journey.

The first leadership book I picked up was the *21 irrefutable laws of leadership* by John Maxwell. I had watched videos and listened to audios of John Maxwell beforehand, and he was constantly talking about teaching. He said that when we learn new leadership principles, we should teach them to our people. So, I did. I started teaching what I learned to my team, and other teams within my engineering depot. But Siemens is a huge, worldwide conglomerate and I wanted to increase my influence with as many Siemens people as possible.

Siemens has an internal social networking site (much like Facebook), and they also have their own video platform (much like YouTube). So, I started to create my own videos based on the *21 irrefutable laws of leadership and* uploading them to the Siemens social network. A lot of people enjoyed them and learned from them. I loved this so the more leadership books I read, the more videos I created.

I am a lifetime student of leadership, personal growth, and influence. By being a student, and constantly learning gives me the capacity to constantly teach and create my own resources, hence this website.

The leadership principles you are learning here must be applied with your people, and then be taught to your people. The other leadership books, courses, videos, or any other leadership content you consume, you must do the same too. That is how we keep the leadership legacy going. That is how you build your own legacy too. Being remembered as the person who improved people's lives through leadership is an amazing legacy to have.

When teaching your people about leadership, it is up to you what you decide to teach them. It's likely that your people have never been taught leadership before, so everything will be brand new to them. That is what makes teaching leadership exciting, you can make this material your own, but only after you have learned and applied it yourself.

When we are teaching our people, we are *showing* them that we care about them and that we respect them. We are putting in the extra time and effort to help them improve themselves and get closer to reaching their full potential. But, before you start teaching any material, we must set our vision with the team. What are you and the team striving for by you teaching, and them learning about leadership? What is the common goal that you want to achieve? That must come from you.

My vision when teaching any of my people, and my vision when writing books or creating other materials is to create highly effective leaders, who will then go on to create more highly effective leaders. My vision is like a snowball effect. That is how I want to increase my influence and leave my legacy.

It will take some time to become the most effective leadership teacher to your team you can be, but as John Maxwell says, *"Leadership is practiced daily, not in a day."* Think of it like going to the gym. Simon Sinek says, *"If you go the gym and work out for 9 hours, you will not get into shape. However, if you work out for 20 minutes every day, you are guaranteed to get into shape, we just don't know when."* So, don't give up on teaching your people because you will be letting them down.

When you set your vision for the team, and what you want to achieve by teaching them how to be leaders, and apply leadership principles to their lives, they will be inspired to do everything they can to strive towards your vision. It may be a vision that you may not ever achieve, but if you do this right, then your team will work their hardest to try to achieve it. That's why having a vision is very powerful. It brings people together and is something that is common between the whole team. Your vision eventually becomes their vision.

For example, when I began creating videos for Siemens and teaching my team what I had learned, the vision I set for them was simple. I wanted them to become leaders, not just engineers. I wanted them to take what they had learned and teach other Siemens' people too. I didn't want my teaching to stop with them, or my videos, or that I was the only person doing the teaching. I wanted to inspire everybody to learn about leadership, and then pass on what they have learned. That vision to me is beautiful. Creating a world of highly effective leaders is who I am.

I really hope you understand my vision, and my purpose, and I need you to make it part of your purpose and vision too. Commit to passing on what you learn in leadership. Commit to teaching your people, so that they can teach their people. Not just in the workplace, but with their family, friends, and anybody else they know. Do it through teaching face to face, videos, books, or any other resources that you can create yourself. Imagine a world full of highly effective leaders. Imagine having highly effective leaders as our world leaders who thought like you and I did.

To me that is a beautiful world to live in. What do you think?

When setting your vision, what does it look like? Can you see it? Do you believe it? If you cannot, then your vision is fake. It is just words. You must be able to see your vision. That is why it is called VISION.

You must be passionate about it, think about it all the time, and show your team the vision and the dream before they buy-in.

Ask your team the right questions

A curious mindset is the best mindset to have, as opposed to a certain mindset. When you are curious, you are constantly learning. When you are certain, you can't go anywhere else. Curiosity over certainty, any day of the week.

When a highly effective leader asks questions, they either want to learn something new for themselves, or they want to inspire their team to think differently and for themselves. So, when you go back to your team and you begin asking questions, always have those two things in mind. You are not telling anybody what to do, you either want to learn, or you want to inspire self-thinking.

How do you do that? You may be asking. If you are seeking to learn new knowledge from your team then ask *"why?"* If you want to help your team to think differently and for themselves, then ask *"how?"* Highly effective leaders are very skilled at asking these types of questions, and they are very deliberate with this technique. They either have a goal to learn, or a goal to help.

When asking the question of *"how?"* this can be very powerful when a new change is being implemented either within your team, or within the organisation. Asking *"how?"* enables you to be proactive and embrace the change. Rather than reactive and fear the change. Most are reactive when it comes to change, so being proactive will put you and your team in the minority. However, it will put you and your team in a better position, and people will start to take notice. Especially your immediate senior leaders, and other senior leaders within your organisation.

As the leader of your team, reacting negatively to change, and fearing change will influence your team to react and behave in the same way. So, you are negatively influencing, rather than positively influencing. This is something you need to avoid. Leaders who react negatively to change are looking for the cons of the change, and don't even consider the pros. Even if the pros outweigh the cons, a leader who fears change will still focus on the cons. When leaders behave in this way, they and their team will become a victim, rather than the recipient of change.

When you become the victim of any change, it will feel that the change is happening to you, rather than for you. When change happens to you, you will not have any control over the change, and neither will the team. Your team will feel like they have been manipulated, and if you cannot show leadership then this will create distrust. If you are not thinking positively about this new change, then your negativity will flow into other aspects of your work.

Highly effective leaders embrace change, and they are proactive to change rather than reactive. When a highly effective leader embraces change, they are also embracing the responsibility that comes with it. By embracing, and accepting the responsibility, that is how they are proactive to the change. Usually, when a new change is being brought in, people ask the obvious question of, *"What is happening?"* However, a highly effective leader's first question to a new change begins with *"How?"* When asking *"How?"* you are seeking the solution from the very beginning, which is a lot more powerful than asking *"What?"*

Many times, in my career, when a new change is about to come into the organisation, you always hear the same questions from the negative people. Whether they are the leaders, or the team, they are always the same as the following:

- What benefit am I going to get out of this change?
- What benefit is our team going to get out of this change?
- What help am I going to get with this change?
- What help is my team going to get with this change?
- Who is going to make this change happen?
- Will I need to do anything to make this change happen?
- Will my team need to do anything to make this change happen?

These types of questions are a reaction to a new change and is a very negative response.

When a highly effective leader and their team embrace a new change, they turn the questions around, and will ask the following:

- How can I benefit this change?
- How can my team benefit this change?
- How can I help others benefit with this change?
- How can my team help others benefit with this change?
- How can I help make this change happen?
- How can my team help make this change happen?

As you can see, just putting the word *"How?"* at the beginning of the questions, gives a whole different perspective on yours and your team's way of thinking. It completely changes your way of thinking from negative and reactive, to positive and proactive.

A new change usually means a different way of being and behaving. So, we must accept the responsibility that comes with this new way of being. Leveraging the question of *"How can I?"* or *"How can we?"* rather than, *"Can I?"* or *"Can we?"* is how highly effective leaders embrace and accept the responsibility and lead themselves and their team through positive change.

Highly effective leaders have no doubt that the change they are leading is the right thing to do.

When a low performing leader asks the question, *"Can I?"* they are doubting themselves already and being negative. There is no substance to the question *"Can I?"* so you don't know whether you can, or you can't.

However, when a highly effective leader asks, *"How can I?"* they are making a commitment and have made the decision that they can make this change happen, no matter what. When this commitment is made, the only thinking you will do will be in a positive direction, and if you come up against any obstacles, you will find the solution to remove them. The word *"How?"* puts you and your team on the right path.

When you are thinking positively about any type of change, whether that be in work or at home, you will always find a solution. There may be more than one solution, but you will think of them all and decide on the best one. Just by using the word how. Teach your team to do the same thing, and they will start thinking in this way too.

Low performing leaders will always be the victim to change, and because they play the victim, they will feel stuck. It is like they are stuck with the change, instead of embracing the change. When a highly effective leader embraces the change, they will feel unstuck and creative. They see the change as a benefit for all and do everything they can to ensure the change is seen through to the end.

When it comes to a new change being brought in for your team, department, or organisation, then ask yourself the *"How can I?"* and *"How can we?"* questions above. See for yourself how different you will be thinking, and how different you feel. You will feel a lot more positive and accepting of the new change.

Then, once you have asked yourself, and the team has asked themselves these positive questions, ensure you take positive action. Don't just ask the questions and do nothing. Take positive action and see what happens.

You and your team will start to devise solutions for obstacles, rather than just coming up with obstacles that you don't do anything about. Train your team to put *"How?"* in front of *"Can I?"* and feel the difference from being reactive to proactive.

To get buy-in, you will get further by asking questions. If you make a statement, then there is nowhere to explore.

Make decisions as a team

Thinking for ourselves is how we really know what our next step to success is. Asking questions is how we help others to think for themselves so that they can do the same.

When I begin to lead new team, it fills me with so much joy. It gives me the opportunity to lead more people, help more people, and spread my message even further. There is nothing better than leading a team through a transformation or a change and seeing them grow throughout the process. My process is to ask the team a lot of open questions, so that I dig deep into their minds and get them thinking deeply. Some of the questions I ask, I already know what they are going to say. However, sometimes I have no idea.

I love it when I ask my team a question that I think I know they answer to, and they correct me with either a better answer, or prove that I was completely wrong. It shows me that they are thinking deeply, and differently because they are not expecting me to just give them the answers. If I was to just give them the answers instead of asking the team questions, then the team wouldn't need to do any thinking, and even if they didn't agree with my answers, they would not challenge them. Which would tell me that I do not have buy-in from the team. So, to make life easier for myself, I ask the team questions that they need to come up with the answer for. If they come up with the answer, then they have already bought-in. whereas if they are just told what to do, it would be a lot harder for me to get the team's buy-in.

When the team are coming up with the answers and the solutions, everyone is bough-in, and everyone is involved in the decision-making process. When you decide with your team, and value everyone's input, the team will feel respected.

There are a lot of low performing leaders who think that it is their role to make all the decisions and come up with all the answers. For a lot of these leaders, they have not had the leadership development they have needed so they can learn that it is not their role at all to provide all the answers and make all the decisions. These are not bad people, or have bad intentions, they just don't know. I was one of these leaders. I thought that if I made all the decisions and had all the answers then I would be helping my team. But I was pulling the team back and depriving them of the opportunity to think for themselves.

There are other low performing leaders who like to come up with all the answers because of their ego. They like to look like the leader who knows everything. They think it makes them look good when it doesn't.

Low performing leaders who make all the decisions, give all the directions, and come up with all the answers may change things for a short time. But long term, leading like that will only hinder the team and their progress. By not making the team think for themselves, or even think at all is pulling them back. Low performing leaders do not think long term, they only really look for quick wins. They say they think long term, but they really don't.

A lot of low performing leaders cannot see past today. The furthest that a low performing leader will think is next week. That is not long-term thinking at all. They do not have their team's best interests at heart, even though they say they do. Giving the team the answers that they should be thinking of themselves does not help them, it hinders them.

Highly effective leaders are always thinking long term. They are "big picture" thinkers. Their long-term thinking is focussed on their people, not on the product or service that the organisation provides. Highly effective leaders know that when they are asking their team open questions, and getting them thinking deeply, they are increasing their influence with them. They know this because they are giving influence on their team, which in turn increases their own influence. Give influence on a person, and you will receive influence. Give to receive.

What you must realise and put a stop to today is, thinking for others. Whether that be your team, your colleagues, your friends, or your family. By answering people's questions for them and doing their thinking for them does not help them at all. It pulls them back, rather than propels them forward. Helping people to think for themselves is how you help them to grow, and how you do that is by asking them open questions. Getting them to think deeply, and come up with their own solutions, ideas, and make their own decisions.

Asking open questions is thought provoking for your team member. It will create a spark in them to start thinking for themselves, and they will need to spend the time trying to come to the answer or the solution. So, let them spend that time thinking, and don't rush them.

A great way to challenge yourself when you are leading your team is, only lead them with questions. Do your very, very best to not give your team any directions whatsoever. The only answers you can give them to their questions is an open question. Do this for about a week and see the difference in your team when they are forced to start thinking for themselves.

It is difficult to pull off when leading your team with questions only because you feel the urge to just give them the answer. Especially if it is just a small thing, they are asking you about. But, believe me, leading with questions works. It will get both you and the team into new habits.

You will get into the habit of asking questions, and intentionally helping your team to think for themselves. It will become second nature to you. Your team will get into the habit of thinking for themselves, and they will reduce the number of questions they ask you daily.

When I was engineering production manager in Scotland, I tried this technique when we used to have our one o'clock catch-up meetings. Every shift we used to have a one o'clock catch-up meeting to see how the day was going, no matter if it was dayshift or nightshift. Usually this was the time when I would get bombarded with questions. But for a couple of weeks, I made sure I answered their questions with open questions. I couldn't believe the difference in how the team were thinking.

They were a lot more prone to coming up with their own ideas without needing my permission, and our one o'clock catch-up meetings became more of a discussion and joint decision-making process, rather than me having all the answers.

You could "feel" the difference in the team, and I could feel the difference in all of us when we were together. Asking thought provoking, open questions to the team brought us together a lot more, and we were making decisions together. If anything went wrong, I was accountable. When things went right, the team got the credit. That's how I saw it, and the team felt respected because of it.

Leading your team with questions is how you make decisions as a team. When you make decisions as a team, the team feels a lot better. The team are more willing to listen to each other and help each other. You as the leader will find it a lot easier to gain the team's buy-in, and as a leader this is all you need to make things happen.

A LIMITLESS MINDSET

A simple open question can lead to something amazing.

Delegation is not telling people what to do and how to do it. Delegation is telling people what is to be achieved

CHAPTER 8
LEAD THE TEAM THROUGH PROBLEM SOLVING

Grow your team through delegating

A highly effective leader does not make all the decisions. They understand that making all the decisions for the team is not leadership, it is management. We do not manage people, we lead people.

When you delegate to your team, what is it you delegate? My instinct when I started delegating was to delegate what to do, and how to do it. Basically, telling them what to do. However, what I wasn't realising was, by delegating in that way, I was still making myself responsible. The reason for that is, if the task went wrong, then the team member would have an excuse of, *"Well, you told me to do it like that."* However, the whole point of delegation is to make others responsible. How do we do that? We delegate what we want to achieve, not what to do. We leave it up to our team member to figure out how to achieve it, and that will take the responsibility off us.

Highly effective leaders are always looking for ways to grow their team, and delegating responsibility is one of the ways they do this. They use delegation to build on the relationships they already have with their people. The delegate responsibility with the purpose of growing their people and giving them the opportunity to achieve for the team.

When you delegate the "what to do, and how to do it" to your team, you are managing them, not leading them. Most leaders all over the world do not delegate in the right way. They tell their people what to do and how to do it, without knowing that that is the wrong way to delegate. Which of course is not their own fault as they don't know what they don't know.

Low performing leaders delegate by telling their people what to do and how to do it. I know a few low performing leaders who delegate this way because they think it helps them to be authoritative. When in fact they come across in a very egotistical way. Delegation is not about that. Delegation is about allowing a person to take ownership of a task or problem, and growing through achieving the task, or solving the problem. Highly effective leaders use delegation as an opportunity to develop their people and grow their team.

When a low performing leader delegates, and tells their team what to do, and how to do it, he/she is creating a team of people who cannot think for themselves and will become highly reliant on being told what to do. The leader is not empowering their people with any responsibility, they are not growing their people, and they are not showing their people any respect. So, when their team member finishes a task, he/she will just wait until being told what to do next. They will only tell the leader they have completed the task if the leader tells them to tell them that.

This is unfortunately very common, especially in my area of work which has been engineering. Telling people what to do is a very quick way of delegating, and it is very short term. There is no consideration for the long term, and there is not any "big picture" thinking. A highly effective leader will avoid telling their people what to do because they want them to think for themselves, not rely on them to tell the team what to do.

I worked with a leader who told me to come to him whenever I had finished a task. The team and I were not allowed to go any further until we had consulted him, and he told us what to do next. I felt disrespected, and so did the rest of the team. We were constantly told what to do and needed his "say so" every time we wanted to move on. I didn't feel as if I was growing or developing. We were busy for busy sake and were working on autopilot when we were given our tasks.

The leader did all of our thinking for us, we didn't have any responsibility, and we didn't feel like we owned any task or problem. The leader owned it all.

The team and I would often talk about not being allowed to move on, and we would have our own recommendations for next steps when we had finished a task. So, we did have the ability to think for ourselves. If this leader were to allow us to get to the end of a task, and then recommend the next step, then I am certain we would have felt a lot better. We wouldn't feel as disrespected, and we would have been given the opportunity to grow.

When a leader gives their team the opportunity to recommend the next steps, they are transferring the responsibility, and the thinking over to the team. They are putting their trust in their team and building trust with them rather than creating distrust.

The role of the leader now changes, and they become more of a support rather than the decision maker. The leader is supporting their team in their development and growth. The leader is also thinking longer term and looking at the "big picture" in a lot more detail.

Nowadays, when I delegate to the team, I am leading, I tell them what we need to achieve, and I leave it up to the team to work it out. I will always be there for support when necessary, but I do my best to avoid telling the team what to do. I like the team to report back to me when they have completed the task, and that can be verbally or in a formal report. I love it when they report back, and I see their eyes light up as they are talking with passion. That is when I know they have grown and developed.

A highly effective leader will transfer full responsibility over to their team when delegating. They put their complete trust in their team, and in return the team trusts them. The team expect to take full responsibility and ownership and will very rarely question the leader's decision in delegation. A highly effective leader puts their full confidence in the team to make the correct decisions as a team.

As a highly effective leader, it will be your responsibility to delegate effectively. That is how you share responsibility with your team, it is your team's responsibility to achieve the desired result, it is your responsibility to delegate, grow, and develop your team.

It is our responsibilities in the workplace that determine our role. Not what our job description says.

When there's a problem, don't solve it for the team

Solving your team's problems for them automatically makes you responsible for the problem. Asking them questions about the problem that will help them to solve it, automatically makes them responsible.

I like to talk quite a bit about asking questions of your team, rather than giving them the answers, and solving their problems for them. The reason for that is questions is what a highly effective leader relies on. They use questions as a tool to get their message across to their people. They use questions as a tool to gain buy-in from their people. They use questions as a tool to increase their influence with their people.

When I learned and decided to stop trying to solve problems for my team members, I always had the same question to whatever their question was. My first question was always, *"What's the problem?"* My team would get confused by that question because they think they have just told me. But, asking *"What's the problem?"* made them think again and dig a little deeper. I wouldn't stop asking that question until they had thought a little deeper and come back to me with the *actual* problem.

A lot of the time, my question should have been, *"Who is the problem?"* because the problem did usually start with the person seeking help. The only real problem was, they didn't know that they were the problem. And they didn't know that to solve the problem, they had to think a lot deeper than what they were thinking, or if they were thinking at all.

After a while, the team knew that if they came to me with a problem, I would ask them the same question every time, *"What's the problem?"* So, they did their thinking beforehand, and when they did that, they realised that they didn't need to come to me anymore because they had already solved the problem.

I must admit, I did find it quite fun when the team would come to me with problems, and I would ask them, *"What's the problem?"* because I would go and observe them thinking deeper about the situation, and it gave me a sense of pride. It made me feel good that I had prompted my team to think a lot deeper and solve their own problems. It was a very simple, but very effective thing to do.

When I was a leader of a team in the railway, we were responsible for solving train problems, especially when they broke down in service. So, if I saw a team member trying to solve a train problem while sitting at his desk, I would change my question from, *"What's the problem?"* to *"Where's the problem?"* If they were going to get to the root cause of an engineering problem on a train, wouldn't it be better to be exactly where the problem was?

When I asked the team that question, they knew immediately to leave the office and go to *where* the problem was. I used to do this myself when I was a young engineer, but in those days, it was rare you could get out of the office because the manager always wanted to keep his eye on us.

However, a highly effective leader is very visual in how he/she thinks. That is why casting a vision of the future for the team is so important. They also know how important it is to *see* a problem, not just rely on spreadsheets, data, and information being told or sent to them. Whenever I got the opportunity to see a problem, I knew I had a much better chance of solving it.

So, if you have a team member who is responsible for a problem, then you need to give them the best chance they can at solving it. That means letting them go to the problem. Don't keep them in the office so you can keep your eye on them. Give them the freedom, and the tools to solve their own problems, wherever they may be.

If you don't give them this freedom and opportunity to solve their problems, then you will ultimately be responsible because you are responsible for your team member.

I liked to ask my team that when they think they have solved a problem, they should report back to me with a report or presentation. This would give me the opportunity to ask more questions, and get my team thinking even deeper, and assure me that the team have done everything they could. The type of question I would ask during a presentation are, *"How do you know that the problem is solved?"* If they could answer that question well, then I was satisfied. However, if they answered me with, *"I just know."* Then that would not be acceptable. Especially if they had *seen* the problem and they were just relying on information passed to them by somebody else.

There are far too many big decisions being made in organisations that are based on poor data, poor information, and hearsay. They are the main problems in a lot of organisations throughout the world. Big decisions need to be made together as a team and based on *seeing* and *solving* problems in the right way. We cannot make decisions based on gut feelings, or information we found in the system from a year ago.

I know it takes a lot more time to give your team the opportunity to solve problems correctly. But take it from me, it takes a lot more time when you decide on something based on poor information, and you must start all over again or backtrack.

Asking questions until you cannot ask anymore is the only way you can be 100% sure everything has been done to solve a problem.

As I have said before, when you ask your team questions, and you are helping them to solve their own problems, you are showing your team respect. They will feel a lot more valued and will follow your lead by asking each other questions when they are collaborating on problems. When collaborating on problems, and asking questions of each other, they are showing each other respect too.

Earlier, I told you that one of my old managers wouldn't let me go to *see* problems on trains in service because he wanted to keep his eye on us. Well, this same manager would often ask us questions too. He would often ask, *"Why are you doing that Tom?"* or *"Who told you to work on that?"* Whenever he would ask those questions, I would feel nervous or intimidated because what he meant was, *"What you are doing is wrong."* Or *"I don't want you to work on that."* So, they were asking the wrong questions.

When I started to ask questions, I had a completely different outlook. I thought like a child did. Why? You might be asking. Well, if a child asks you, *"Why are you doing that?"* they do not want you to stop what you are doing, they want to know so that they can learn from you. That is exactly the frame of mind I have. I do want to learn from my team. I have no interest in getting them to stop what they're doing or telling them that they are doing it wrong.

When I explain to my team *why* I am asking them, *"Why are you doing that in that way?"* they are a lot more relaxed and open to explaining it to me. They know that I do genuinely want to learn from them.

It's a great feeling when you begin to see your team asking their teammates the same kinds of questions, and because they genuinely want to learn. They do not want to stop them from doing anything or blame them in anyway.

Problems are there to be solved. But how do you solve a problem? Should you first see the problem before you can solve it? Or should you work from information and data that you got from somebody else? It's your choice.

Develop your leaders

If you could create leaders, who believe what you believe, follow you because they want to, and have the passion to create leaders themselves, imagine what kind of organisation you could build.

The more you show respect for your people, the more it will get noticed by other leaders within your organisation. Some of the leaders (low performing leaders) will react negatively to you, but the other leaders will react positively. Even the senior leaders will notice you and will consider you an asset who can help them develop the other leaders and their teams within the organisation. I have had this happen to me before when I worked at Siemens, and I was asked to create resources that would help develop some of the leaders.

It is highly likely that the organisation you work in, whether it is your own business or not, when it comes to leadership development of your people, a leadership expert will be hired. Most organisations throughout the world will not use their own people when it comes to developing their leaders. The reason is, they do not encourage their people to work on themselves every day in the area of leadership, just like I am doing with my books.

To be a highly effective leader, you should not wait for your organisation to hire in a leadership expert to train you for a week and then disappear. Leadership is a lifelong journey for you, and it is a lifelong journey for your people too.

Most leaders do not pay any interest in leadership development, as they don't believe they need it. They have the position of leader, and that is all they need. That is how most low performing leaders think. However, because you work on yourself every day, and you are striving to become a highly effective leader, you are in a great position to take the responsibility for leadership development in your organisation. Just like I did when I worked for Siemens in London. The resources I produced still help leaders in Siemens today and they were delivered in 2015.

That doesn't mean that you need to start creating leadership resources, but it does mean that you must take responsibility for developing leaders, starting with your own team.

You should position yourself as the leadership development resource for your organisation. You have been reading, studying, and applying leadership principles in your own life, so you are in a great position to help others within your team and the organisation to do the same. Being the leadership development resource will also help you to keep learning and developing your own leadership and personal growth. By continuing to work on yourself every day, helping others will form part of that work you are doing.

"Leadership is not about you, but it does start with you." That is one of my favourite leadership quotes, and it is one of the most prominent. If you are not developing your own leadership, then how can you help others to develop theirs? As you are teaching your team, and others the leadership principles that you have developed yourself on, you will be continually learning as you do it. Your people and you will be learning together and helping each other along the way.

The more people you help develop into leaders, the more your influence will increase, and that is the key to being a leadership resource for your organisation. Leadership is influence, so leverage the leaders that you help develop. As they increase their influence with others, your influence will flow through them.

The leaders at the same level in the organisation as you will be asking for your help too, as you help develop the people who report into them. This is a great opportunity to increase your influence with leaders who already have a lot of influence within the organisation, and the industry that you are part of. Build strong relationships with them, help them, and they will help you in return. They will use their influence to raise you up and help you to increase your influence.

As you are helping their people to develop too, you are reducing problems that they currently have in their team. By helping their people to develop their character, and strive to become a highly effective leader too, the number of followers you will have will be increasing day by day. The leaders of the organisation will be happy to follow you because they want to.

I would recommend to you that you create leadership materials. For me, I created 3–5-minute videos on a leadership topic. I took a leadership book, I read through it thoroughly and learned everything I possibly could. I would apply what I learned in my own life first, and then I would create leadership videos from the content. If there were ten chapters in the book, then I would create ten videos based on each chapter. I would try my best to make them 3-5 minutes as this seemed to be the best for the people I was helping.

This might seem a bit strange to start creating leadership materials, but I assure you, they really will help you and others. Especially when increasing your influence throughout your organisation. You are on a leadership journey, and so is everybody else within your team, and the other teams. Travel on the journey together.

When I created leadership videos at Siemens, I received so much support from the senior leaders, including the CEO of Siemens UK at the time, Jürgen Maier. The most influential leaders within your organisation will support you too, and your resources will become more in demand each week. You will be viewed differently, and others will want to follow you in creating their own materials. Whether that be in leadership or other topics. It will be great to see others want to follow in your footsteps and help others.

However, at first, you will get a bit of resistance from some people, and they will not buy-in to you straight away. Don't let this phase you, just keep going with what you are striving to achieve. Keep practicing what you are learning from my books, and the other leadership and personal growth resources you are studying. Keep working on yourself every day and keep helping others to do the same. Take the responsibility for teaching others and create your resources. Make it your mission to become the leadership development resource of your organisation, and I promise, you will get there.

As you become the leadership resource for your organisation, your team will want to help you. They will want to become a leadership resource too. Other leaders will want to become leadership resources. So, before you know it, you will have built a team, and a culture of leadership development, and leadership resources throughout the organisation.

Focus on leading and developing leaders. Help them to develop themselves first, then help them to help others develop. We are not talking positions or titles here. We are talking about passion and the desire to help each other.

To achieve something good, do it alone. To achieve something great, do it with others. Leadership is about others, not about you.

Know what your team are capable of, and know what you want for them

CHAPTER 9
UNLEASH THE TEAM WITH LEADERSHIP

Embrace your team's potential

Highly effective leaders know how to develop their team to be the best people, and team they can be. Low performing leaders don't know what the team is capable of, or what is required to make them the best they can be.

Before you embrace your team's potential, you must first embrace your own potential. As a leader of a team, we all want our people (or we should all want our people) to be the best they can be and go the extra mile for us and the team. However, if we are expecting this of our team, then we should be expecting the same from ourselves and leading by example.

What we need to realise is, we are in exactly the right place we should be at this moment in time. The reason we are in exactly the right place we should be is because of us, nobody else. You are the reason you are where you are. I am the reason I am where I am. If you want to change your current circumstances, then you must first start with yourself. If you feel that the results you are getting should be better, then you need to better yourself. Nobody else can do it for you.

How is your team performing at present? If they are under performing, then you are not leading them right. A highly effective leader serves his/her team, and a low performing leader actually believes it should be them to be served. If you are not currently serving your team, and they are under performing, then you need to change your leadership style to one of service to your team.

When I ask you to *"serve your team"*, what I really mean is help your team.

When I have used the term serve in the past, a lot of the engineers I worked with didn't like the term. So, I stopped using it and just used the term help. Help and serve both mean the same thing. To lead a team, what you are really doing is helping them. What you are helping them to do is to perform to their best, be the best person they can be, and complete their work to the best of their ability.

It should be your purpose to influence every member of your team, so that they can increase their influence with each other, and others within the organisation. Highly effective leaders know that leadership is influence, and that everybody within their team has influence, so therefore everybody has the potential to be a leader. You just need to embrace that leadership potential you have in every member of your team.

You have a team of leaders already; they just lead on a lower level to you.

You lead on a different level to your own senior leaders. Followers only follow leaders who are on a higher leadership level to what they are. For example, your senior leaders will not see you as their leader, and usually won't follow you, so you will not be able to develop their leadership skills. However, that does not mean that you cannot influence them.

Who you are (character) is what determines what level you are in your leadership. What you know (competency) plays a part too, but a much smaller part. For example, when you interview a person to join your team, you are a lot more likely to hire them because of who they are, not what they know. I have hired people who were not as qualified as some of the other candidates, but they knew who they were, and I had that feeling.

We have spoken previously that leadership is 80% character and 20% competency. If there is a member of your team that you cannot connect with, then what and how much they know is irrelevant. You are much more likely to achieve good results with your team, if you can build a meaningful relationship with them. If there is no trust between you and your team, then they will not give everything they have, and be the best they can be.

For you and your team to achieve the best results possible, you must focus on strengthening your own character, and building strong relationships with each member of your team. Both as individuals, and together as a team. Creating a relationship of trust is what's most important. With trust you will find it much easier to gain buy-in, and the team will want to do their best, and be their best for you and everyone on the team.

As a highly effective leader, it should be your mission to embrace your team's potential, and this means developing them as leaders. When you have built strong relationships, and the team are performing well in their day-today tasks, then you must now go that extra mile for them. Moving beyond the job position and title and developing the team into leaders is how you continue to build those strong relationships.

Your team will follow you because they want to. They will feel the importance of working on and developing their leadership. They will understand how to influence others as they are becoming highly effective leaders, just like you.

Teaching and developing your team in their leadership is what will be remembered. When doing this you are positioning you and the team to be the most successful you can be. When you are in this position, you have the potential to achieve the best results you can.

We discussed leadership levels earlier. As you are developing your team into leaders, your leadership level will also be raising higher. As you and your team's leadership levels increase, your results will continue to improve. So, the answer to the question, *"How do we keep improving our results?"* is, raise your leadership level.

As you are going through this book, and putting into practice what you are learning, you are automatically raising your leadership level. As you teach your people what you are learning from this book, your team's leadership level will be raising, as will yours even more.

So, think to yourself and discuss with your team, what results do we want to see? Do we want to have good results, great results, or amazing results? You know what you need to do, just keep raising your own leadership level, and help your team to raise their leadership level too. Highly effective leaders strive to achieve amazing results with their teams.

To achieve the best results possible, you need to go further than what is required. You and the team need to go the extra mile for each other, and for the organisation. But first, you need to become the best leader you can be, and your team need to become the best leaders they can be. If you keep working to your best, you will achieve the best results possible.

However, for you to have a team of leaders, who are the best people they can be, you need to keep investing in them and developing them. As I said earlier, they will feel your investment, and they will feel different when they are being developed. It is the feeling that will be remembered, not just the fact that you have invested in them. It is this feeling that the team will want more of, and that you will also want more of. So, to keep getting that feeling, keep developing.

Having a leadership position is one thing. Moving beyond the position is another. Developing your people to be the best they can be is the only thing.

Work on weaknesses but focus on strengths

The stronger we become as leaders, the stronger team we can build. The stronger team we can build, the more we can achieve together.

Everybody has strengths and weaknesses. Most people focus on their weaknesses and neglect their strengths, so after a while their strengths become weaker. We should focus on our strengths and develop our weaknesses. Our character is an area that most need to develop and what I would consider a weak area. By developing our character to become strong, our leadership develops at the same time.

A team is not a team until a highly effective leader turns them into a team. The quicker you can do this the closer you will be to becoming a highly effective leader. Low performing leaders do not actually lead teams, they lead groups of people but call them teams.

When you first take over a team, you are taking over a group. They may have worked together with a previous leader and called themselves a team, but it will be up to you as the new leader to turn the group into *your team*. When I have taken over groups in the past, the first thing I discovered was, the people who wanted to be there, and the people who didn't. Some of the groups I had taken over had never even met each other before. So right from the beginning I knew it would be tougher to turn the group into a team.

The most important thing when taking over a new group that you want to turn into a team is, respect. If you want to turn them from a group into a team, then you must show the group respect right from the beginning. When you show respect right from the beginning, you will find it a lot easier to build relationships with the group. You will find it a lot easier to build the group into a strong team. To do this you must work on and build your character to be the strongest it can be. It takes character to turn a group into a team. It takes character to show respect. It takes character to build relationships.

Having worked in the engineering and rail industry for over 20 years, I have seen many engineering managers take over new groups, and really struggle to turn them into a team. The problem for them was, they knew a lot about the job (competency) and were very experienced engineers. However, they had very little leadership training or experience (character). That is the reason I am writing this book, and my other books. The only way to turn a group into a team is through leadership.

A highly effective leader knows that to turn a group into a team, then they must find out what each team member's strongest competencies are. Then to help them grow and improve, they must focus on these strengths. A highly effective leader also knows that they must find out what each team member's weaknesses are. Then to help them grow and improve, they must develop those weaknesses in their character.

Everybody needs to work on themselves every day to strengthen their character, including you. It will be your team member's character (who they are) as the reason they are not where they and the team should be or need to be. When working won your team member's strengths, in the area of competency (what they know), this is what will push them and the team forward.

Highly effective leaders know to work on themselves every day on their character. When it comes to their people, they help them to work on themselves and focus on building their character every day too. A highly effective leader has helped so many of their people with their character, that character building forms their leadership development and personal growth initiatives. Some of their people don't work on their character every day, and so should use these initiatives to change that habit.

I focus 20% of my time developing my weaknesses around character, and 80% of my time working on my strengths around competency when working on myself every day. This also applies when it comes to focussing on the team. I focus 20% of the time developing my team's weaknesses around character, and I Focus 80% of my time working on my team's strengths around competency.

Developing your team's character will help to mitigate most of the problems that occur with team's. Most problems are due to a person's character, not their competency. So, developing character will help reduce these problems.

Most leaders spend 100% of their time working on competency, and do not help to develop their team's character. When they don't develop their own, or the team's character then there is no leadership within the team. Especially when the team are going through a difficult time.

When leaders do not work on, and develop their own character, they do not lead themselves. If they do not lead themselves, then how can they lead a team, or a group of people? When leadership is lacking within a group, it is usually due to the leader's lack of character. If the leader does not develop their own character, then they cannot lead by example and inspire their team to do the same.

As mentioned earlier, when taking over a group of people before you can turn them into a team, you need to know what their strengths are. So, you need to focus on the group's competencies, and to find out their strengths in this area, you need to ask them questions. Questions will generate a discussion. The questions I like to ask are, what do they like to do? What don't they like to do? How long have they been with the organisation? How long have they been in the industry? What was their previous job? Who have they worked for in the past? What are their hobbies? What experience do they have?

These types of questions will spark a conversation between you and the new group member and will allow you to dig deeper.

When you have all the answers you need, you will know how to start building a relationship with the group member and the group. The stronger your relationship becomes, the stronger the team becomes.

It is then that you can start to focus on their strengths and help them to develop their weaknesses. When we focus on our teams' strengths and develop their weaknesses, we are expressing our own humility, and striving towards our team's vision and purpose. When developing our people, we are developing ourselves too. We are becoming stronger leaders for the benefit of our people. We put our people before ourselves, but do not consider ourselves less.

When we are around a leader who is strong and puts their people before themselves, it gives us a great feeling. When they focus on the team's vision and purpose, it inspires us to do the same and follow them.

Being around a humble leader gives us a better chance to achieve what we set out to achieve.

Let the team be the influencers

The more valuable you make your team, the more valuable they will make others.

As your influence increases with your team members, their influence will increase with each other and with other colleagues too. It is your role as a highly effective leader to give your team the freedom to be influencers. As yours and your team's influence increases beyond your usual inner circle, you will find it a lot easier to find extra support to help you achieve your team's mission. You will also find it a lot easier to convince and gain buy-in from colleague's who are not in your team, or even in your department. As your influence increases through your team, they will increase your influence cumulatively, a lot further than you could alone.

With your influence increasing cumulatively through your people, they will be connecting with others, and building strong relationships in the same way as you did with them. As you prioritise personal growth and leadership development with your people, and you keep investing in their growth, yours and their influence will constantly be becoming stronger. The more time you invest in your people, the bigger the benefit will be for both them and you. Investing creates a win-win situation for all.

The more people are influenced by you and your people will want to help you and your team. They will share the same vision as you and will want to work with you to help you and your people strive to achieve it. As they are positively influenced, they will feel the difference both in their professional life, and their personal life. Their mindset will change, and they will want to do everything they can to increase your influence too. Soon, all departments throughout your organisation will have been influenced by you.

This really sounds great, and in theory will be amazing when your influence can be increased cumulatively through your team, then others, and then others beyond them. But how do we do this in practice and continue to work every day to ensure this happens. With leadership development, constantly teach your team principles that will help them to lead themselves and others. As you do this, you are setting them up to increase your influence through them. Then, as you are training them to lead others, teach them principles that go beyond the world of work. As you are beginning to be the best person you can be as a highly effective leader, by teaching your people to do the same, you will receive further support. They will increase their influence with you, and you can then leverage their influence in a positive way.

As you are investing time into your people and going the extra mile for them with teaching them leadership and personal growth, you may receive some negativity from the low performing leaders in your organisation. I received quite a bit of negativity from low performing leaders, they thought I should be "focussing on the job". They are managers, not leaders of people, and they focus on processes, not people. Do your best to avoid this negativity, and don't listen to it.

Whenever I talked to the low performing leaders about why I wanted to invest in my team's leadership development and personal growth, I used to receive some "interesting remarks". For example, *"You're not a trainer, why are you training your team?"* Or *"You're not going to get the work done if you put the team first."* I found these remarks to be very amusing because I knew that by putting the people first, the job would take care of itself. How can you get the work done anyway, if you have a de-motivated team?

Even though I ignore the remarks I receive from low performing leaders, I would try my best to explain to them the reasons why I invest in my people and teach them leadership. It would be better for them and the organisation if more leaders followed me in investing in their people. If we have more and more leaders teaching their team's leadership and personal growth, then imagine what kind of culture we would be creating. Imagine how far our influence could increase. Low performing leaders only see things short term, but if we could convince them to change their mindset and think longer term, then our influence will compound.

The more people who follow us because they want, not because they have to, our influence will increase so far that we will be influencing people who we have never met. We can do that because the people who our people influence will be influenced by us too. Thinking long term and having your vision at the forefront of yours and your people's minds is what you need to focus on if we want to keep gaining buy-in from people.

For example, if I wanted help from an engineer who worked in a different team and had a different leader, it would depend on the relationship I had with his/her leader. As the engineer doesn't report to me, I wouldn't be able to go straight to them, I would have to ask permission. If I had a poor relationship with his/her leader, then they would probably say something like, *"I don't have time to let my engineer do some work for you at the minute. Maybe soon."*

By telling me that there is no time to let the engineer help me, is giving them a way out. There is always time to help each other, but because of the poor relationship, the leader is choosing not to help me. I used to find this quite a lot when I took up my first leadership positions. I didn't dedicate time to building strong relationships in the beginning, so I didn't receive the co-operation I was looking for.

However, instead of just focussing on building relationships with leaders, I began to build strong relationships with engineers outside of my own team. I did this so that I could increase my influence further, and so they could increase my influence outside of my usual inner circle. So, when it came time to ask for the engineer to help me, I would find it a lot easier.

The reason I would find it easier is, the engineer was already influenced by me, so he/she would go their leader and gain their buy-in to supporting me and my team. The leader may not have been influenced directly by me, but their team member would use my influence to gain their buy-in. Just as I had with them, and my own team members. Leveraging other people's influence is a great way to get the support you need from outside your usual inner circle.

When your team, and others in your organisation know your vision, follow your vision, and want to support you, you will find that they will offer you help even when you don't ask for it. There is a saying, *"It's who you know, not what you know."* When it comes to increasing your influence far beyond yourself and your people, that saying could not be truer.

The more people you can influence, the better for you and for your people. As you influence others, listen to them. Embrace their ideas, lead them, and give them the freedom to influence. Allow yourself to be influenced by them.

Directing people is not leadership. Thought provoking questions is leadership

CHAPTER 10
KNOW WHO YOUR LEADERS ARE

Increase yours and your team's influence

Creating change with your people and your organisation is a commitment. You cannot say "I wish this, or I wish that" and expect things to change. When you make a commitment to change, you must say "I will do this, or I will do that." What commitment to change are you going to make?

When in a conversation with your team, do you ask them questions, or do you do most of the talking? What kind of questions do you ask? Open or closed questions? Highly effective leaders know that by asking a member of their team a question is a lot more powerful than giving them a direction. So, to make a more powerful impact with the individual they ask lots of questions.

Being part of teams that were managed and not led, we got asked lots of questions too by the manager. But they were different types of questions. They were not thought-provoking questions that would influence you to think deeply. Managers of people only seek the right amount of information needed for them to make their decision and then give a direction. They are not interested in listening to your ideas on the subject, or any recommendations that you may have.

Highly effective leaders ask lots of questions because they want the team to think deeply about the subject, and in the end make the decision or come up with the solution. The only direction the highly effective leader gives is through the types of questions they ask, open questions (Who, What, Where, Why, How). Thought provoking questions. Highly effective leaders don't want to make the decisions for the team. If they did then the team would not grow, they would stay stagnant. The only way to help the team grow is by helping them to think and act for themselves, and asking great, though provoking, open questions is an excellent way to do that.

When you ask your people questions, you are building a stronger relationship with them, and building trust. It shows them that you care about them, and that you value their thoughts, ideas, and opinions. You are *showing* them respect. Not just talking at them in a one-way conversation. It is a two-way street. Your people, or person will feel closer to you, and feel part of the team. It becomes a "we" situation, not "us and them".

When you ask your people questions, you are opening yourself up. You are letting your guard down and inviting your people to influence you. If they can influence you, you are *showing* them that you have listened and that you understand what they are telling you, what they are feeling, and that you are there for them. If you allow yourself to be influenced, then your influence will increase too, and you will find it a lot easier to gain buy-in and build trust with your people.

When you ask your people questions, you are allowing your team to take ownership. When they take ownership of the situation, they are far more likely to work with you and their teammates to come to a solution and decide. When the team are involved in making the decisions, they feel like more of a team, and you will receive the respect in return that you are giving to them.

When you ask your people questions, you are digging deep into the minds of your people. They are opening to you by thinking deeply, and when you dig, you will start to understand their minds and how they think. Do they have a leader's mindset? Do they have the potential for a highly effective leader's mindset? When you find that out, you will know which members of your team have the capacity and are willing to lead right now, and who still needs to work on it. When you have this information, it will be far easier to know who to focus on and make a priority.

When you ask your people questions, you are empowering them. Empowering is an overused term in leadership, and a lot of the time for the wrong reason. Empowering means that you are sharing the responsibility with your people. You are inviting their opinions, ideas, and recommendations. You are encouraging decision making. You are not dictating to them; you are empowering them. Asking questions is a great way to empower someone.

When you ask your people questions, you are giving them the chance to open up about their frustrations, and what they have on their minds. Most people keep their frustrations to themselves. But, when you ask questions, the person finds it easier to let go of their frustrations and talk about them. This gives you the opportunity to set things straight if what they are thinking is incorrect. Or it gives you the opportunity to help them change how they feel about a certain issue. But, most of all it gives you the opportunity to listen.

When you ask your people questions, you are creating a mastermind. A mastermind is a group of minds that are on the same wavelength, not just one. The leader cannot think of everything by themselves (like my situation as performance manager with Siemens). When they understand this, and invite the team's minds too, they will understand that many minds are far better than one mind. Creating a mastermind environment gives everybody the opportunity to contribute and give their opinions, ideas, solutions, and recommendations. When this happens, the team feels like a synergy. It creates an environment of freedom to express yourself, open, and be yourself.

Creating a mastermind group within your own team is a great thing to do. I have done it, and the feeling you get from it is amazing. Especially when everybody contributes. This environment encourages discussion, teamwork, and allows everyone to empathise with each other. It brings the team closer together. When you ask questions of the mastermind group, you can see your team growing and developing as they answer the questions. If they get stuck, then another member of the mastermind will help. It is wonderful to see. I would encourage you to do the same with your team.

Asking lots and lots of questions also gives you the opportunity to grow and develop. The more questions you ask, the better your questions will become. The more questions you ask, the deeper you can dig into the minds of your people and create highly effective leaders within them. The more questions you ask, the team will give you better answers, better solutions, and ultimately better decision making.

If you ask a poor question, then you will get a poor answer. If you ask a good question, then you will get a good answer. If you as a great question, you will get a great answer.

Always keep in mind, asking questions doesn't mean you do most of the talking. By asking questions, you are giving yourself the opportunity to listen, not talk. We still need to listen 80% of the time. The questions we ask are giving our people the opportunity to speak, not you. Always remember that, and always remember you can ask more than one question in a conversation. The greater questions you ask, the greater answers you will receive.

Low performing leaders answer questions with very short answers that gives their people just the right amount of information. Or, in my experience, most of the time gives you no information at all. They just want to get rid of your question because they don't know how to answer it. However, a highly effective leader will answer a question from their people with a question. They know that if they give them the answer, they are not growing the person. But if they ask a question in return of the question then they are helping their person grow by getting them to think for themselves. Thought provoking questions is an excellent way to grow your people.

By accepting responsibility for the future of yourself, and your people, you will have the power to create the opportunities you need, to build the future you see in front of you. What does the future look like? What is your vision?

Lead by example

Do you want your team to achieve better results, or do you want your team to live better lives?

Making the decision to become a highly effective leader and making the decision to become a leadership resource for the leaders within your organisation tells me a lot about you. This is a huge responsibility to take on and requires a person with a very strong character. You will already have your team members following you because they want to, and you have achieved that by building strong relationships with them and helping them to develop their character. So, for the leaders to follow you because they want to, you need to build strong relationships with them, and help them develop their character too.

Highly effective leaders do not use their title or position to gain power over people, instead they use them as a leverage to help them develop and grow their people, and beyond. Highly effective leaders will purposely follow aspiring leaders, or leaders who are willing to improve, so that they can help them develop. With a highly effective leader following you because they want to, it really does give you the confidence to lead others. You will know this because a highly effective leader is trustworthy, they are authentic, they are transparent, and they have very good intentions.

How committed are you to becoming a highly effective leader? It is not easy at all, in fact it is very uncomfortable, and you will be out of your comfort zone a lot of the time. I have talked about being comfortable with being uncomfortable in my other resources, and to be a highly effective leader, you really do need to get comfortable with being uncomfortable. That is how you are growing. If you are not uncomfortable, then you are not growing, you are static. The more you become comfortable with being uncomfortable, the more your confidence to lead others will increase.

To be able to lead others, and help them raise their leadership level, you must first raise your own leadership level. As a highly effective leader, we work on ourselves every day to keep raising our leadership level, it is a commitment we make, and we honour. Are you willing to make that commitment? I hope you are.

As people start learning about highly effective leadership, and what it means to be a highly effective leader, they tend to look outwards at other leaders. They do this and judge what they are learning on leadership against the leaders that they work with, or their senior leaders, and even the leaders of the organisation they work with. People very rarely look inwards at themselves and judge themselves against what they are learning. Highly effective leaders judge themselves; they do not judge others. You should do the same.

However, even though I am advising you to look inwards, judge yourself against what you are learning, and only be in competition with yourself, it is human nature to look outwards at other leaders. You will do it, and I will do it. It's just how people are. So, if that is the case, with our people constantly looking at us and judging us, we must ensure that what they see is what we are teaching them. If we are behaving in a certain way as a leader, but we are teaching our people something different, then we will come across inauthentic, and untrustworthy. With a lack of trust from our people, we will lose all our credibility. It is impossible to become a highly effective leader without credibility or trust.

When you are teaching your people leadership, you are teaching them a way to live, not a theory, or a problem that needs to be solved. We cannot teach people a way to live, and then live a completely different way to what we teach. We must practice what we preach. We must believe in what we teach and be passionate about what we teach. How you live is what you will be judged on, not what you teach.

When I began teaching leadership, and striving to become a highly effective leader, I needed to make some changes within myself first. I needed to lead by example and lead myself first before I could lead others. If I said I was going to do something to help another person, then I followed through and did it. Previously, I had the tendency to let others down. So, that was a huge thing that I needed to change in myself. There were a few other things too.

You may want to make similar changes in your life so that you can live what you teach. I will be able to help you with this through my other books and materials.

You will be teaching people how to build their character and build relationships with others to increase their influence and strive to become a highly effective leader. So, you must do no less. In fact, you must do more if you want to set that example for your leaders. I don't mean you have to live a perfect life. But you must certainly be working hard every day to live the same way you are teaching others to live.

As you are teaching your people to work on and strengthen their character, you will begin to see weaknesses in their character that you can help them with. What you must not let happen is, your people seeing weaknesses in your character. That will also lose your credibility.

As you lead your people by example, you are *showing* them what highly effective leadership looks like in real life. You are not just teaching the theory of it. As your people *see* what highly effective leadership looks like, they will have a model that they can follow. As they begin to model your behaviour, they will start to feel different, people will start behaving differently around them, and they will start to be treated differently. Your example will set a clear direction for your people to go, and they will know that following your lead, will enable them to be the role model for others in the same way.

As the leader teaching others through the resources you create, you will be giving your people something in common. You will be giving them a way to live and behave both in their personal and professional lives. They will then be able to pass on this commonality to their people. The commonality being you. It is your character that enables you to behave like a highly effective leader. It is your behaviour that your people will see. It is what your people see that will either attract them to you or not. Make your behaviour attractive to your people, so that they will want to model that behaviour. Give your people your vision that they can make their own and believe what you believe.

We must never just expect our team and our people to just know how to behave. It is a low performing leader who thinks this way. That is how they get out of the responsibility of leading by example. If the team do not have a vision to strive towards, believe what you believe, and a model of how to behave, then your team are not going to model highly effective leadership.

Highly effective leaders accept the responsibility of highly effective leadership. The responsibility of highly effective leadership is to create an environment for their people that will allow them to learn, grow, develop, and follow a role model. What kind of environment are you going to create?

If people follow leaders because they must, then there is no freedom to choose. If people follow leaders because they want to, then there is freedom to choose. Are you going to allow your people the freedom to choose?

You've achieved success, how do you sustain it?

Investing in your processes is a good thing, for the short term. Investing in your people is a great thing, for the short term and the long term. Think wisely when deciding what to invest time and money into.

So, now that we know how to teach our leaders, and we know how to help them to become highly effective leaders, how do we help them sustain and put into practice what we are teaching, or have taught them? One of the biggest challenges that I find when teaching a person, a leadership principle is, how do I help them put into practice what they have learned, every day? I don't want to phone them up every day to make sure they are working on themselves, that is micro-managing them.

What we need to do as leaders is ensure that we do what we say we are going to do. We want to make leadership development and people development a priority, so let's ensure that we do that. Otherwise, what we teach, or what we role model will not be put into practice.

As I am writing this last section of the book, I am hoping that you are willing to look within yourself and be honest with yourself. Do you really want to become a highly effective leader? Do you really want to help others to do the same? Are you willing to work on yourself, every day?

Take some time to reflect on who you are, and what type of leader you are. Take some time to think about what type of leader, and what type of person you want to become. I am hoping that you want to become the kind of leader who focusses on their team, and their people who surround them. I am hoping that you want to increase your influence with your team, and your people. I am hoping that you want to have respect for your team, and for your people every day. Your team, and your people need you to be that person, and be that leader.

Your organisation, your industry, your friends and family, and the entire world need you to become a highly effective leader to lead them and show them the way. They need you to show them that they too can become highly effective leaders and lead the highly effective leadership movement that you can create.

Work on yourself every day, study leadership and personal growth, and teach your people to do the same. If you can do that you will have started a highly effective leadership movement, and you will help create more highly effective leaders.

I hope you can feel the passion I have for helping create a world of highly effective leaders. I hope you share the passion that together, along with many others can help create that world.

Imagine a world of highly effective leaders who work on themselves every day to build their character, so that they can influence others to do the same.

If you can imagine that world then take what you are learning from me and pass it on to your people. They are craving our help.

When teaching your team, and the people around you, keep in mind that you are doing this to help them become the best person and the best leader they can be. You are not doing this to improve the bottom line of the organisation or improve the results of the team. They are by-products of improving the people. Take care of the people first, and the results, and numbers will take care of themselves.

We are respecting our people. We are not respecting our figures, numbers, reports, processes etc.

Our return on investment is an organisation of highly effective leaders who want to increase their influence to help create more highly effective leaders.

On the other hand, high impact leaders never mention ROI. They know as I do, it doesn't matter. They do it because of their character, not because of the ROI. If there is truly respect for the people, you don't invest in the development of the people for the ROI. You do it because they matter.

I really hope you feel what I am writing on these pages. If you can feel the words and they are inspiring you, then that is when you know you are in the presence of highly effective leadership. If the words mean nothing to you, then keep searching for words that do mean something to you.

Quite often when I look at vision and mission statements on the walls of organisations, they don't mean anything to me. The reason for that is a statement is not enough. We need to be *shown* vision. That is why it is called a vision, we need to *see* it, and we need to *feel* it.

A LIMITLESS MINDSET

If organisations do not invest the profits they make in their people, then they are not going to develop their people into the best people and leaders they can be. If they don't have their people at their best, then they are not going to produce the best results for the organisation, and the organisation will not achieve the most profits possible. It is a vicious circle, but it is extremely simple. Invest most of your money and most of your time into your people, and you will produce the best people, the best leaders, the best results, the most profits, and the most fulfilled organisation possible.

Organisational leaders need to take the courage and change how they see their business. Business is not about resources, processes, products, profits. Business is about people. Without your best people, you have no resources, processes, products, or profits.

As a highly effective leader, it is your mission to help create a team, and an organisation of highly effective leaders. How do you do that? You work on yourself every day to be the best person and leader you can be. You teach your people to do the same. You live what you teach and practice what you preach.

If you can do that with your senior leaders, then you can leverage them to invest time and money into the people, so you have more leverage to help them become the best people and leaders they can be. If you can create an organisation of highly effective leaders, imagine what kind of organisation you could have.

Value your people, and you will have more valuable people. Care for your people, and they will care for their people. Invest in your people, and they will invest in their people. Respect your people, and they will respect their people.

It's not about the results, and it's not about the profits. It's about the people. Take care of your people, and the results, and profits will take care of themselves.

It would be great to hear your thoughts on how this book has helped you as you climb the leadership ladder. I would also love to know how you are doing by helping others to climb the leadership ladder, and beyond.

So, please feel free to email me and share your thoughts at:

tom@highlyeffectiveleader.com

To order my books, online courses and my other resources, please visit www.highlyeffectiveleader.com or www.amazon.com

Highly effective leadership cannot only be taught. It must be LIVED

ABOUT THE AUTHOR

There are hundreds of books that teach us how to be a "leader". Most of them are good and teach the right things. But, Tom Lawrence has seen very few that teach us how to influence our people, or take our personal growth seriously, or even lead if you're not in a "leadership position".

Tom wants to change that by writing books for people who are interested in leading and inspiring others.

He wants to help people who are not in a leadership position but are aspiring to become a leader, current managers who want to take the next step and become highly effective leader, and to help senior leaders enhance their leadership skills.

Tom began his career in 1999 as an apprentice mechanical engineer, working for an automotive company in Liverpool, UK.

After completing his apprenticeship he was made redundant from that company and had to find other work, that is when he joined the rail industry working for Merseyside's train operator. He worked there for six years, and during that time he achieved his degree in mechanical engineering, his master's degree in maintenance engineering, and he started his first leadership role as a project manager in 2009.

After leaving Liverpool in 2011, he has worked in Edinburgh, Glasgow and London where he currently lives. Tom became a chartered engineer in 2013, and is a mentor for new and upcoming engineers working towards their chartership. Working in these different cities he had remained within the rail industry and led different types of engineering teams, learning and practicing different leadership styles.

Tom has vowed to make it his life's work to help you become successful. He hopes the information that is provided through his books, goes some way towards helping you achieve all you ever wanted from your career as a leader.

Tom's Leadership Mantra

Learn to <u>Lead</u>, <u>Grow</u> and Increase your <u>Influence</u>

Tom's Why

To develop leaders and aspiring leaders into <u>highly effective leaders</u> so that, they can develop leaders and aspiring leaders into <u>highly effective leaders</u>.

Tom's Vision

Tom can see a world in which our leaders create environments that focusses on people and helps them to learn to <u>lead</u>, <u>grow</u>, and increase their <u>influence</u> for the good of us.

Made in the USA
Coppell, TX
02 February 2022